STUDENT LABORATORY WORKBOOK

MEGAN MOCKO
University of Florida

MARIA RIPOL
University of Florida

STATISTICS: THE ART AND SCIENCE OF LEARNING FROM DATA
THIRD EDITION

Alan Agresti
University of Florida

Christine Franklin
University of Georgia

PEARSON

Boston Columbus Indianapolis New York San Francisco Upper Saddle River
Amsterdam Cape Town Dubai London Madrid Milan Munich Paris Montreal Toronto
Delhi Mexico City Sao Paulo Sydney Hong Kong Seoul Singapore Taipei Tokyo

The author and publisher of this book have used their best efforts in preparing this book. These efforts include the development, research, and testing of the theories and programs to determine their effectiveness. The author and publisher make no warranty of any kind, expressed or implied, with regard to these programs or the documentation contained in this book. The author and publisher shall not be liable in any event for incidental or consequential damages in connection with, or arising out of, the furnishing, performance, or use of these programs.

Reproduced by Pearson from electronic files supplied by the author.

ISBN-13: 978-0-321-78342-4
ISBN-10: 0-321-78342-5

1 2 3 4 5 6 BRR 16 15 14 13 12

www.pearsonhighered.com

PEARSON

Table of Contents

Preface

Introductory statistics courses can be challenging for students who may be intimidated by the material even before starting the class. It can also be challenging for the instructor, particularly when teaching courses where students' mathematical backgrounds differ greatly. We find that the best approach is to teach our classes in a way that involves students in the learning process as much as possible. This workbook is what we use to teach our classes. It is divided into two parts: lecture notes to be completed in the classroom and activities that can be used in a variety of class settings. Both the lecture notes and activities are designed to promote active learning by the students.

Lecture Notes

The first portion of the workbook includes lecture notes for the first ten chapters of the Agresti/Franklin, *The Art and Science of Learning from Data,* 3rd edition, textbook. These lecture notes are a shell to be completed during class, where students have all the important information at their fingertips allowing them to concentrate on understanding the theoretical concepts rather than copying them down. More importantly, it also allows for examples to be completed during class with the students as active participants as opposed to getting all the answers without thinking about them.

Activities

The back of the workbook contains 20 activities that can each be finished in 50 minutes or less. Each of them is designed to have students learn statistics by doing statistics and gives them the opportunity to explore the theoretical concepts and the applied procedures in more depth.

These activities are very flexible: they can be used in a standalone laboratory section associated with the class, they can be incorporated into the lectures in a small class, and in some cases they can also be used as homework assignments. The activities accompany chapters 1 through 11 in Agresti/Franklin, *The Art and Science of Learning from Data,* 3rd edition, and they supplement the concepts discussed in the textbook and the lecture notes. The activities are printed on perforated pages that can be detached and turned in for grading.

Some of the activities give students first-hand experience with the entire statistical analysis process: generating and gathering the data, entering it in the computer, analyzing it, and presenting the results and their interpretation. Other activities use data sets, applets, and computer simulations, which allow students to gain a much better understanding of important statistical concepts such as the Central Limit Theorem, or the ideas behind a confidence interval. Applets can be found on the CD that accompanies all new copies of textbook and data sets can be found on the CD and at www.pearsonhighered.com/mathstatsresources.

On the next couple of pages you will find a short description of each of the activities, including information about which sections in the book they correspond to and what you need for each of them. Some of the activities can be completed individually while others require collecting data, which makes them ideal in the classroom as a group activity, but can also be done by students outside of class. Some of the activities require the use of a computer for simulations or accessing the Internet, other activities would find a computer helpful, but not necessary, in order to analyze data and a few activities don't require a computer at all.

Activity	section	G=group I=individual	Computer needed?
Activity 1: Introduction to the General Social Survey What is the General Social Survey? Explore the General Social Survey website and see real data firsthand.	any	I	Yes
Activity 2: Describing Data with Dotplots So, who are your fellow classmates? In this activity you will collect data on your classmates, describe the data collected and explore some of the challenges of collecting data.	2.2	G	No
Activity 3: Hurricane Season What is a typical hurricane season? Use graphical and descriptive summaries to describe different hurricane seasons.	2.2 – 2.5	I	Yes
Activity 4: Exercise and Drinking Habits What is considered a normal amount of exercise or a typical drinking habit for a student at your university or college? Computing conditional probabilities will help you investigate this behavior.	3.1	G	No
Activity 5: Understanding Correlation Get a grasp on correlation. Use applets to explore the properties of the correlation coefficient.	3.2	G	Yes
Activity 6: Foot Size, Forearm and Height Is there a linear relationship between the lengths of someone's foot size and forearm length or their forearm length and their height? Use statistical software or a calculator to help you explore these relationships.	3.3	G	Helpful
Activity 7: Sampling Legos® What is the big deal about a simple random sample? In this activity you will explore three different types of sampling methods and determine what works best overall.	4.2	G	Helpful
Activity 8: Should You Trust Everything You Read? You shouldn't trust everything that you read. Take a look at a news article provided by your instructor and look at it through the lenses of a statistician.	4.1 – 4.4	I	No
Activity 9: Normal Rent Payment Is the rent that you pay higher, lower or about the same as other students? Explore rent payments in your local area. Is the variable rent normally distributed?	6.2	G	Helpful
Activity 10: Sampling Distribution of a Proportion What is a sampling distribution? This activity allows you to explore the shape of the sampling distribution of the sample proportion and to discover when this distribution is approximately normally distributed.	7.1	I	Yes
Activity 11: Central Limit Theorem How do sample means vary from one sample to another? This activity allows you to explore the shape of the sampling distribution of the sample mean and to discover when this distribution is approximately normally distributed.	7.2	I	Yes

Activity	section	G=group I=individual	Computer needed?
Activity 12: Confidence Interval for a Proportion Have you ever wondered what proportion of Americans share a certain opinion? Use real data to make a confidence interval for the population proportion of Americans who feel a certain way using the General Social Survey.	8.2	I	Yes
Activity 13: CI for the Mean – What Does it Mean? How much time does a typical student at your campus study? In this activity, you will collect data from the class to investigate how much time students at your campus spend studying as well as to practice identifying correct and incorrect interpretations of the confidence interval.	8.3	G	Helpful
Activity 14: Testing for Proportions Based on hypothesis statements of your choosing, you will conduct a significance test for a proportion based on data from the General Social Survey.	9.2	I	Yes
Activity 15: Interpreting P-Values How much time do students at your college spend watching television? Investigate this by performing a significance test for a mean based on data that you collect.	9.3	G	Helpful
Activity 16: Two Scoops of Raisins Are there really two scoops of raisins in every box? In this activity, you will investigate a manufacturer's claim that they put two scoops of raisins in each box of cereal.	9.3	G	Helpful
Activity 17: Haircuts – Who Spends More? Who spends more per haircut, men or women? What about per year? Investigate this using a statistical software package or a calculator.	10.2	G	Helpful
Activity 18: Matched Pairs Does listening to soothing music really lower your heart rate? Explore this claim using the statistical inference techniques that you have learned.	10.4	G	Helpful
Activity 19: Comparing Two Groups Have you ever wanted to write your own survey? This activity allows you to be the creator of the survey and to analyze the data that you collect from the survey.	10.1 - 10.4	G	Helpful
Activity 20: Contingency Tables One of the most frequent examples of data that we see in real life is data in a contingency table. This activity explores different concepts related to contingency tables.	11.2	I	Helpful

Chapter 1 Statistics: The Art and Science of Learning from Data

1.1 Using Data to Answer Statistical Questions

What Is Statistics?
The art and science of learning from data – designing studies, analyzing the data, and translating data into knowledge and understanding of the world around us.

Who Uses Statistics?
Everyone.
- Marketing –

- Medical Studies –

- Government –

- Social Scientists –

- Environmental Studies –

- Engineering –

- Sports –

Why Use Statistical Methods?
- **Design** – plan how to obtain the data (Ch 4).
- **Description** – summarize the data with graphs and numerical summaries (Ch 2-3).
- **Inference** – using data from a **random** and **representative sample** to draw conclusions about the **population** of interest. We need to understand some probability concepts to do statistical inference – this allows us to attach a measure of reliability to our inference or conclusion. (We will learn about probability in Ch 4-5 and about statistical inference in Ch 8, 9 and 10.)

1.2 Sample vs. Population

Data sets consist of:
Subjects - persons, animals, or objects in our study / experiment.
Variables - the characteristics that we measure on each subject. We call them variables because they can take on different values for each individual.

Population – all subjects of interest.
Sample – subjects for whom we have data.

Random Sampling – each member of the population has the same chance of being included in the sample. Random samples tend to be representative of the population, so we can draw better conclusions about it.

Parameters – numerical summary of the **population**.
Statistics – numerical summary of the **sample**.

Example: Internet sites report that about 13% of Americans are left-handed. Is this true for students at your university? During a statistics exam, the instructor walks around the room and counts 15 left-handed students out of 98 students in the class.

a) Identify the following:

   Variable:

   Population:

   Sample:

   Parameter:

   Statistic:

b) Was random sampling used?

Chapter 2 Exploring Data with Graphs and Numerical Summaries

2.1 Different Types of Data

Categorical Variables – place each observation into groups and they are usually summarized by the **percentage** of observations in each group.

Quantitative Variables– take on numerical values. Key features are the center (**average**) and spread (**variability**) of the data. Quantitative variables can be further split up into:
- **Discrete** – take only a finite list of possible outcomes, such as a count (0, 1, 2, 3, etc.)
- **Continuous** – has an infinite list of possible values that form an interval, even though sometimes we are limited in our ability to measure them

Example: The class list below has all the information for each student in a class at the end of the semester, including their year in school, major, exam grades, project grades, number of absences, their average in the class and their final letter grade in the class.

Student ID#	Name	Yr	Major	Ex1	Ex2	Pr1	Pr2	Abs	Avg	Grade
46895382	Aiken, John	1	Psych	78	82	20	24	2	81.6	B
21657845	Bailey, Kim	2	PolSci	62	74	15	19	10	68.0	D
13695544	Carr, May	2	BusAdm	95	92	25	24	0	94.4	A
	…etc.									

Which of the previous variables are:

    Categorical   Discrete Quantitative   Continuous Quantitative

2.2 Graphical Summaries of Data

The type of graph used depends on the type of variable.
Most graphs are done with a computer, particularly for large data sets.

Graphs for Categorical Variables: Bar Charts and Pie Charts

Example: Year in school for students in classroom

	Frequency (Count)	Proportion	Percentage
Freshman			
Sophomore			
Junior			
Senior			
Total			

Bar Chart

Pie Chart

Graphs for Quantitative Variables: Dotplots, Stem-and-Leaf Plots, and Histograms

Example: Grades on an exam for a small class:
82, 76, 65, 94, 72, 80, 91, 45, 72, 86, 89

Dotplot

Stemplot

Histogram

For this data set, what can you say about:

a)   the center of the distribution?

b)   the spread of the distribution?

c)   the shape of the distribution?

d)   unusual observations?

Some common shapes:
Mound or Bell-shape

Uniform or Rectangular

Bimodal

Skewed Left

Skewed Right

Examples: For the following graphs, describe their distributions in terms of shape, center, and spread.

1. Graph of starting annual salaries for engineering graduates.

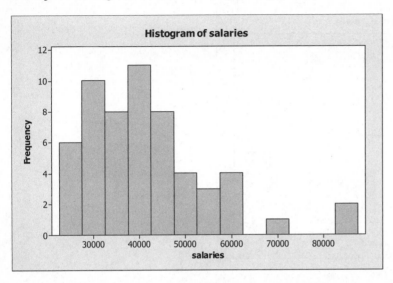

2. What age group has the most dangerous drivers? The following graph represents the number of driver deaths (per 100,000 licensed drivers) by age group, in one year.

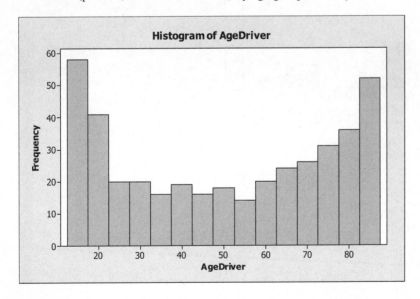

Examples: The following graphs present the population growth, by race, for each state when comparing the numbers from the 2010 census to the 2000 census. The data for each state appears on the table in the next page. Interpret the graphs' findings.

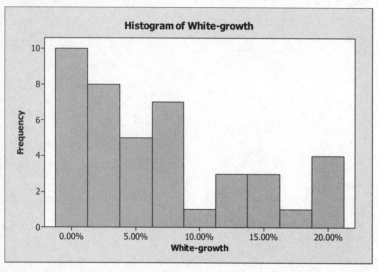

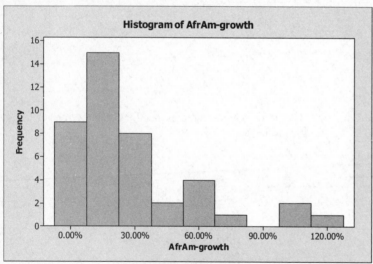

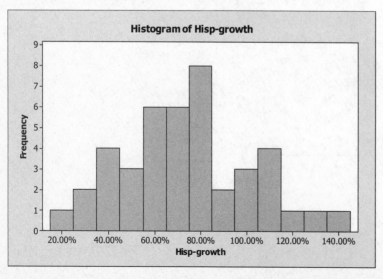

State	White-growth	AfrAm-growth	Hisp-growth	State	White-growth	AfrAm-growth	Hisp-growth
Alabama	3.60%	8.30%	144.80%	Missouri	4.40%	10.20%	79.20%
Alaska	9.00%	6.80%	51.80%	Montana	8.30%	49.60%	58.00%
Arizona	20.50%	63.00%	46.30%	Nebraska	2.60%	20.90%	77.30%
Arkansas	5.00%	7.40%	114.20%	Nevada	19.00%	61.40%	81.90%
California	6.40%	1.60%	27.80%	New Jersey	-1.20%	5.50%	39.20%
Colorado	14.90%	22.20%	41.20%	New Mexico	15.90%	23.90%	24.60%
Connecticut	-0.30%	16.90%	49.60%	North Carolina	12.50%	17.90%	111.10%
Delaware	5.80%	27.30%	96.40%	North Dakota	2.10%	103.30%	73.00%
Florida	13.20%	28.40%	57.40%	Ohio	-1.10%	8.20%	63.40%
Georgia	8.60%	25.60%	96.10%	Oklahoma	3.00%	6.40%	85.20%
Hawaii	14.40%	-2.60%	37.80%	Oregon	8.20%	24.30%	63.50%
Idaho	18.60%	79.80%	73.00%	Pennsylvania	-0.70%	12.50%	82.60%
Illinois	0.60%	-0.60%	32.50%	South Dakota	4.50%	117.90%	102.90%
Indiana	2.80%	16.00%	81.70%	Tennessee	7.90%	13.30%	134.20%
Iowa	1.20%	44.10%	83.70%	Texas	19.60%	23.90%	41.80%
Kansas	3.30%	8.90%	59.40%	Utah	19.40%	65.90%	77.80%
Kentucky	4.60%	14.00%	121.60%	Vermont	1.20%	104.90%	67.30%
Louisiana	-0.70%	0.00%	78.70%	Virginia	7.20%	11.60%	91.70%
Maryland	-0.90%	15.10%	106.50%	Washington	7.80%	26.20%	71.20%
Minnesota	2.80%	59.80%	74.50%	Wisconsin	2.80%	18.00%	74.20%
Mississippi	0.50%	6.20%	105.90%	Wyoming	12.50%	27.60%	58.60%

Example: The following histograms represent use different bin sizes on the x-axis to describe the point differential (points scored by the Gators minus points scored by the opponent) for each game played by the Florida Gators Baseball team during the 2011 season.

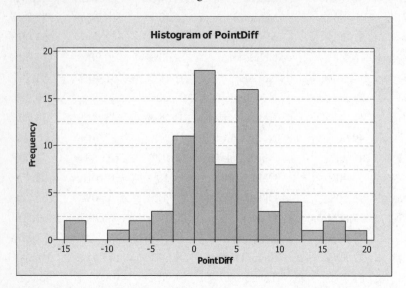

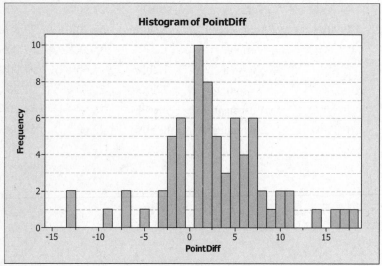

a)   How many games did the Gators play?

b)   How many games did the Gators lose?

c)   How many games resulted in a tie?

d)   How many games did the Gators win by 10 runs or more?

e)   How many games did the Gators lose by 10 runs or more?

2.3 Measuring the Center of Quantitative Data

Measures of Center
- **Mean** – the average of all the observations: $\bar{x}$
- **Median** – observation "right in the middle": M
- **Mode** – most frequently occurring value

Mathematical Definitions:
Notation

n = # of observations in the data set

$x_1, x_2, x_3, \ldots, x_n$ = first, second, third, …, last observation
summation notation (capital sigma)

Formulas:

Mean:

Median:
-
-

Example: Number of "friends" in Facebook (an online social networking service for college students), for a sample of 9 female Facebook members.

288   254   476   329   191   121   404   184   505

a) Find the mean:

b) Find the median:

Example: Number of "friends" in Facebook (an online social networking service for college students), for a sample of 10 male Facebook members.

65   342   345   172   46   458   106   153   244   810

a) Find the mean:

b) Find the median:

Example: The following is a stemplot done in Minitab of self-reported college GPAs of students enrolled in a large introductory statistics course. Find the median of this data set.

Stem-and-Leaf Display: cGPA

```
Stem-and-leaf of cGPA  N  = 114
Leaf Unit = 0.10


    2    2  01
    3    2  2
    5    2  55
    8    2  677
   16    2  88899999
   39    3  0000000000000000011111111
   53    3  22222222223333
  (24)   3  444444444445555555555555
   37    3  6666666667777777777777
   17    3  8888888999
    7    4  0000000
```

Median =

Notes:
- First column represents the cumulative counts from top and bottom.
- The line with the parentheses contains the median.
- Second column contains the stems.
- Rest of the columns contain the leaves.

- Leaf unit:

  0.10   decimal between stem and leaf
          Example from the first row: 2 01 means 2.0 and 2.1

  1.0   no decimal
          Example from the first row: 2 01 means 20 and 21

  10.0   add a zero after the leaf
          Example from the first row: 2 01 means 200 and 210

  0.01   move decimal to left of stem
          Example from the first row: 2 01 means 0.20 and 0.21

2.4 Measuring the Variability of Quantitative Data

Measures of Variability (Also Called Measures of Spread or Dispersion)

Range = maximum – minimum

> **Example**: Find the range of the distributions of the number of "friends" in Facebook for the samples of female and male Facebook members. Data appears below in order.
>
> Female: 121   184   191   254   288   329   404   476   505
>
> Male: 46   65   106   153   172   244   342   345   458   810

Variance and Standard Deviation
Measures of spread around the mean, particularly useful for bell-shaped and symmetric distributions.

> **Variance**:
> - averaged squared deviation from the mean: s^2
> - its units of measurement are those of the original data squared
> - we need to take the square root before we interpret
>
> **Standard Deviation**:
> - square root of the variance: s
> - its units of measurement are the same as those of the original data

Mathematical Definitions:
Recall Notation:
> n = # of observations in the data set
> $x_1, x_2, x_3, \ldots, x_n$ = first, second, third, …, last observation
>   summation notation (capital sigma)

Formulas:
>   **Variance:**

>   **Standard Deviation:**

Why n-1 in the denominator? The denominator in the formula for the variance, n-1, represents the number of degrees of freedom (df). This is the number of independent quantities we are adding up in the numerator. Since $\bar{x}$ is computed from those n observations, only n-1 of the distances from the observations to the mean are independent of each other.

Example: Suppose your whole grade in a class is based on four exams, worth 100 points each. Here are the grades you know:

	Ex 1	Ex2	Ex3	Ex4	Avg
	82	76	85	??	80.5

How many possible values of Exam 4 grade are there?

Examples: Two very simple data sets. For each one, make a quick plot of their distribution, find the mean, median, and range. Compare the two distributions. Then find the standard deviation using the formula, and also using the statistical functions of your calculator.

Data Set 1:   1 1 1 4 7 7 7   **Data Set 2:**   1 3 4 4 4 5 7

Interpreting the Standard Deviation, s:
- The larger the standard deviation, s, the more spread out the data set is.
- s can never be negative.
- s can only be zero if there is no variability in the data – all the observations are identical.
- s is very much affected by outliers.
- s works best for bell-shaped and symmetric distributions.

Empirical Rule - in any bell-shaped and symmetric distribution you will find approximately:
- 68% of the observations within one stdev of the mean.
- 95% of the observations within two stdev of the mean.
- 99.7% of the observations within three stdev of the mean.

Example: One of the many different scales for IQ scores states that the IQ values for the whole population follow a bell-shaped and symmetric distribution with mean 100 points and standard deviation 16 points.

a) Sketch a graph of this distribution.

b) Between what two values will you find the central:

68% of IQs?

95%?

99.7%?

c) What percentage of people have IQs over 148?

Example: Heights of college-aged women are approximately bell-shaped and symmetric, with almost all of them (95%) falling between 4'11" (59") and 5'11"(71").

a) Sketch a graph of this distribution.

b) Find the approximate values of the mean and standard deviation.

2.5 Describing Distributions Using Percentiles and Boxplots

Quartiles divide the data set into four quarters:
- Q1 = Lower Quartile (Q_L)
  = 25% of observations below it = 25th percentile
- Q2 = Median
  = 50% of observations below it = 50th percentile
- Q3 = Upper Quartile (Q_U)
  = 75% of observations below it = 75th percentile

There are many ways to find the quartiles, each resulting in slightly different numbers.
For our purposes,
- Q1 will be the median of the lower half of the data.
- Q3 will be the median of the upper half of the data.

Example: Find the quartiles of the distributions of the number of "friends" in the Facebook for the samples of female and male Facebook members. Data appears below in order.

Female: 121   184   191   254   288   329   404   476   505

Male: 46   65   106   153   172   244   342   345   458   810

InterQuartile Range – measures the spread of the central 50% of the data.
IQR = Q3 – Q1

Five Number Summary of Positions:
Minimum, Q1, Median, Q3, Maximum

Boxplots - graphs based on the five number summary.
- Box contains the central 50% of the data (going from Q1 to Q3). Line crossing the box represents the median. Whiskers extend to the minimum and maximum observations.
- In Minitab (and many books and computer packages) the whiskers only extend to the smallest and largest observations that are not considered outliers. The outliers are plotted individually, with an asterisk if it's a "mild" outlier, and an open circle for an "extreme" one.

Example: Construct boxplots for distributions of the number of "friends" in Facebook for the samples of female and male Facebook members. Compare the distributions in terms of their most important features.

Example: A large statistics course has 18 TAs in charge of the discussion sections. Do final grades in the class depend on who the TA for the section was? The boxplots below show the final percentage grade in the class for all students who took the final exam, by TA in charge of the section.

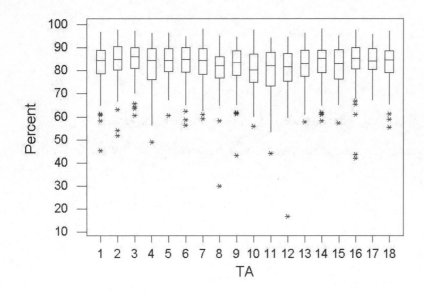

Chapter 3 Association: Contingency, Correlation, and Regression

When studying the association between two variables, we usually want to distinguish between:
- **explanatory** (or predictor) variable
- **response** variable

3.1 The Association between Two Categorical Variables

Contingency Tables:
- Both explanatory and response variables are categorical.
- Display counts (frequencies) on the table.
- Compute percentages to determine association.

Conditional Proportions: Find percentages by dividing each cell count by the total number of observations in their group (as defined by the explanatory variable).

Example: Do male college students follow their school's teams more closely than females? The following data was collected in class on a Monday morning, after a particularly exciting and important basketball game. The question asked was: Did you watch the game on TV last night?

	Whole Game	Part of the Game	None of It	Total
Male	10	12	4	26
Female	21	24	30	75
Total	31	36	34	101

a) What is the explanatory variable?

b) What is the response variable?

c) Find the conditional proportions of each gender that watched all, part, or none of the game.

	Whole Game	Part of the Game	None of It	Total
Male				
Female				

d) Is it fair to say that males were more likely to watch the game than females?

3.2 The Association between Two Quantitative Variables

Scatterplots
- Plot of Y vs. X, two quantitative variables, measured on the same individual
- X = explanatory variable, Y = response variable

Interpreting Scatterplots
- Direction – positive or negative?
- Linear trend? How strong?
- Any outliers?

Examples: Interpret the following scatterplots.

1. X= year and   Y= percentage of Freshmen applicants admitted into UF,
    as reported on the UF website every year by the Office of Institutional Planning and Research.

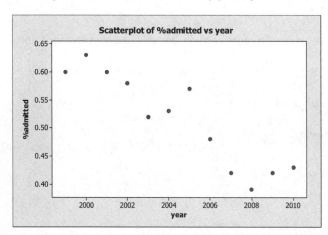

2. Y= percentage of population with Internet access in a country
    X= GDP = country's Gross Domestic Product per capita (in thousands of US dollars)
    Find the US: GDP=47.0 and 77.3% of adults have Internet access   (Agresti/Franklin)

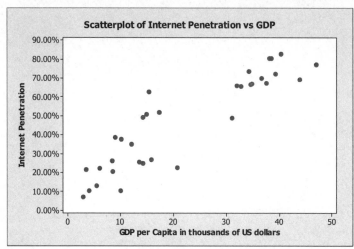

Example: Determine which variable should be explanatory (x) and response (y), and sketch a scatterplot of the relationship you would expect between:

1. father's height and son's height

2. father's height and mother's height

Correlation
- The correlation summarizes the **direction** and **strength** of the straight line relationship between x and y.
- The two variables have the same correlation regardless of which one is called the explanatory or the response variable.
- We will use the symbol **r** to represent the correlation coefficient.
- r is always between -1 and +1 (no units).
- **Interpretation:** positive/negative, strong/weak.
- Outliers can have a strong effect on r.

Some examples:

Formula: Correlation Coefficient
   Note: The formula for the correlation coefficient is mostly for illustration purposes. For homework problems, you can use Minitab, or your calculator (if it does 2-variable statistics).

Why does the formula give the same sign (positive or negative) as the graph?

3.3 Predicting the Outcome of a Variable

Review of Straight Lines:
- All points are exactly on the line
- Line extends forever in both directions
- Equation of a Line:

In Statistics – Regression Line:
- Points are scattered
- Want the equation of the line that best fits through the middle of the points
- Use it to predict the response variable y for a particular value of x.
- We call the predicted values $\hat{y}$.
- Regression Equation: $\hat{y} = a + bx$

Interpreting the Regression Line:
- **b is the slope** (rise/run)
  The slope represents the average (or predicted) change in y for a one-unit change in x.
- **a is the y-intercept** (the point where the regression line crosses the y axis)
  The y-intercept corresponds to the predicted value of y when x=0 and it is necessary to complete the equation. However, we only interpret it if x=0 makes sense and it is close to the values of x observed.

Example: Suppose the regression equation to predict y= weight (in pounds) from x= height (in inches) for female college students is $\hat{y} = -200 + 5x$.

a) Interpret the equation.

b) Predict the weight of a female college student whose height is 5'5" (65").

c) Roughly sketch the relationship between height and weight for college-aged women. Do you expect all 65" tall women to weight the same?

Using Data to Find the Slope and y-Intercept of a Regression Line
There are many lines that, visually, seem to fit a scatterplot well. So how do we find the "best" line to describe a data set? The least squares regression method finds the line that minimizes the prediction errors.

Residuals
- Residuals are the prediction errors for each observation.
- Graphically, they are the vertical distance from the point to the line.
- Residuals = difference between the observed and predicted values of y.
- Residuals = $y - \hat{y}$

Least Squares Regression Method
- The least squares regression line goes through the middle of the points, in the sense that the distances to the points above and below the line cancel each other. That means the sum of the residuals will be zero.
- The least squares regression line also minimizes the sum of squared residuals, or prediction errors. The formulas for slope and intercept are discussed below – they are not hard to derive, but it requires knowledge of calculus.
- Because it is the vertical distances that are minimized, it is important to determine which variable is x and which one is y.
- The least squares regression line passes through the point ($\overline{x}, \overline{y}$).

Least Squares Regression Formulas:

$$\hat{y} =$$

$$b =$$

$$a =$$

R^2: Coefficient of Determination
- $R^2 = (r)^2 = (\text{correlation})^2$
- Easier to interpret than the correlation coefficient.
- It is interpreted as the percent of variability in y explained by the linear regression on x.

Example: After more than 20 years with a 55 mph speed limit, the state of New York raised the speed limit to 65 mph on some of its major highways, effective August 1, 1995. Will highway fatalities increase with the speed limit? Some people say yes, pointing to the reduction in deaths when the 55 mph limit was implemented in 1974. Others say no, pointing to better roads and car equipment (tires, anti-lock brakes, air bags) that help save lives. We will use data on highway fatalities and fatality rates (number of deaths per 100 million vehicle miles traveled) from 1970 to 1993 to do a regression analysis.

a) Why is it better to use fatality rate instead of fatalities?

b) Describe the most important features of the plot of fatality rate vs. year that appears below.

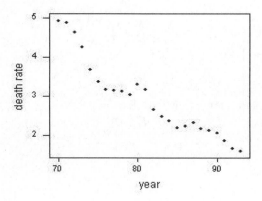

c) Here are some other laws passed during the period being studied. Do they seem to have affected highway fatalities?

    1974 Speed limit reduced to 55 mph
    1981 Anti DWI law, late November
    1982 19 yr old drinking age on Dec 1
    1985 Mandatory seat belt law on Jan 1
    1985 21 yr old drinking age on Dec 1

d) Interpret $R^2 = 90.9\%$

e) Compute the correlation coefficient, r, and interpret.

f) Use the following information to find the regression equation:

	year	death rate
mean	81.5	2.939
stdev	7.07	0.987

g) Plot the regression line on the scatterplot above. Comment on the fit of the regression line.

h) Interpret the slope and the intercept (if appropriate).

i) What fatality rate does the line predict for 1996? Do you trust this prediction? Explain.

j) The fatality rate for 1996 was 1.34. Find the residual for this point and interpret.

k)   What fatality rate does the line predict for 2006? Do you trust this prediction? Explain.

The graph below shows the results of the regression analysis that incorporates data for more recent years.

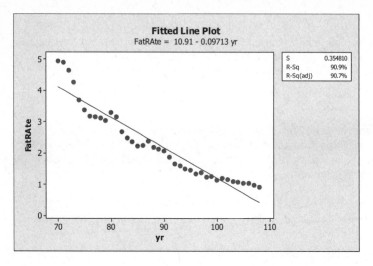

l) How does this regression line compare to the old one?

m) Will the fatality rate ever reach zero in practice? Explain.

n) What fatality rate does this regression line predict for 2006?

o) What fatality rate does this regression line predict for 2015?

p) What could be done to improve our predictions?

q) The plot below shows the results of the quadratic regression. Comment on the fit of this regression curve.

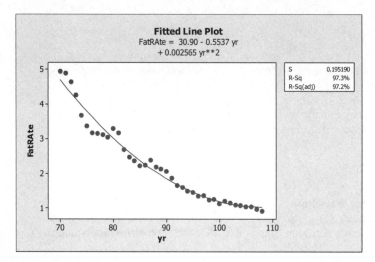

3.4 Cautions in Analyzing Associations

Extrapolation
Predictions made using the regression equation can only be trusted for values of X within the observed range. Predicting outside this range is called **extrapolation**, and you might get ridiculous predictions.
Example: predicting death rate in NY for the year 2020, or even interpreting the intercept in many cases.

Influential Outliers
Outliers in regression are those points that are way away from the trend of the other observations. Influential outliers are only points that have an x value far away from the rest, and that fall far from the trend that the rest of the data follow. These points tend to pull the line towards them, and deleting them can have a huge effect on the regression line, sometimes even changing the sign of the slope.

Example: The following plot has the college GPA and age of several students. Sketch what you expect the least squares regression equation for this data will be, and then do the same thing after removing the 28-year old student. Comment on the difference between the two lines.

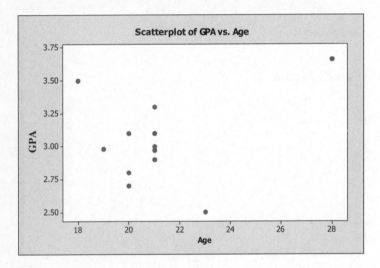

What to do if your data has outliers?

- Check the data and correct any typos.

- If there are still unusual observations, try to find out more about them. Do they belong in the data set? What makes them different? If they do not belong in the data set, you should delete the point before proceeding with the regression analysis.

- If the point is valid, conduct the regression analysis with and without that point. If their results are similar, you may use them. If they are very different, you should collect more data to find out the true relationship between x and y.

Correlation (or Association) Does Not Imply Causation
In many studies, the goal is to prove that changes in x cause changes in y. However, even strong association does not imply causation. It is very hard to prove causation, since there are usually other **lurking variables** that can affect the relationship between x and y.

Examples:

1. Average number of hysterectomies performed by male doctors is much higher than for female doctors. Are male doctors insensitive to women's concerns?
    *   explanatory variable:
    *   response variable:
    *   graphical summary:
    *   potential lurking variables:

    *   conclusion:

2. Newspaper report: Decaf drinkers have higher blood pressure levels than regular coffee or non-coffee drinkers. Is decaf bad for your health?
    *   explanatory variable:
    *   response variable:
    *   graphical summary:
    *   potential lurking variables:

    *   conclusion:

3. Very high positive correlation is found between the size of the head of school-aged children and their reading skills. Are big-headed kids smarter?
    *   explanatory variable:
    *   response variable:
    *   graphical summary:
    *   potential lurking variables:

    *   conclusion:

4. "Fidgeters" are found to burn more calories than calm people. Should we all start fidgeting to lose weight?
    *   explanatory variable:
    *   response variable:
    *   graphical summary:
    *   potential lurking variables:

    *   conclusion:

5. Thousands of studies over the years have found that smokers have a much higher incidence of lung cancer than non-smokers. Does smoking cause cancer?
    *   explanatory variable:
    *   response variable:
    *   graphical summary:
    *   potential lurking variables:

    *   conclusion:

Simpson's Paradox
The direction of an association between two categorical variables can be reversed if we include a third variable and re-analyze the data. This is known as Simpson's Paradox.

Example: A study in the United Kingdom asked women if they smoked or not, and twenty years later determined if they were alive or not. Data appears on the table below. (Agresti/Franklin).

Smoker	Dead	Alive	Total
Yes	139	443	582
No	230	502	732

a) Compute the percentage of smokers and non-smokers who died. What does the data suggest about the effects of smoking? Is this surprising?

b) A potential lurking variable in this study is the age of the women at the beginning of the study. This information was also available, and has been included in the table below. Compute the percentage of smokers and non-smokers who died for each age group. What does the data suggest about the effects of smoking now?

Smoker	18-34 yrs old		35-54 yrs old		55-64 yrs old		65+ yrs old	
	Dead	Alive	Dead	Alive	Dead	Alive	Dead	Alive
Yes	5	174	41	198	51	64	42	7
No	6	213	19	180	40	81	165	28

Chapter 4 Gathering Data

In chapters 2 and 3, we have learned to look at data and summarize it, which is sometimes called **exploratory data analysis**. We haven't made a formal attempt to generalize what these data showed us to larger groups. That is the purpose of statistical inference, and it is crucial then that our data represents the larger group.

Statistical inference makes a statement about the population, based on a random and representative sample and includes a measure of how confident we are in the statement. In this chapter we learn how to gather good data to answer the question of interest. In Chapters 5, 6 and 7, we learn the behavior of data that comes from a random sample. In Chapters 8, 9 and 10 we learn how to do statistical inference.

4.1 Experimental and Observational Studies

Experiments: the researcher assigns subjects to certain experimental treatments.

Observational Studies: researcher does nothing to the subjects but observe x and y.

Advantage of Experiments over Observational Studies:
It is very difficult to determine causation from observational studies, because of possible confounding effects between explanatory and lurking variables. Two variables are confounded when their effects on a response variable cannot be distinguished from each other. Unfortunately, sometimes it is unethical, or very difficult to conduct experiments. Even observational studies, when done properly, can provide good data.

Where do we find the data:
- Anecdotal evidence – not good
- Census: Official government data from the library or the Internet
- Take a sample - If done right, it can be as good as asking everyone, but faster and cheaper. However, we need to take the fact that we do not have data for the whole population into account when presenting our results.

Example:
A researcher wants to know how long it takes adults to master inline skating, and if there are differences due to gender and age. Comment on the following proposals:

a) She could find people who are inline skating in a park and ask them how long it took them to learn, how old they were at the time, and their gender.

b) She could find 10 women in their twenties and 10 men in their fifties, all of whom have never tried inline skating. They are all taught how to do it. Time to mastery of the sport is recorded.

c) She could find several women and several men in different age groups, all of whom are interested in learning inline skating. Each person is allowed to choose the way they will learn: on their own, with friends, or through a class. Time to mastery of the sport is recorded.

d) She could find several women and several men in different age groups, all of whom have never tried inline skating. Have them all taught by the same person, in the same setting. Time to mastery of the sport is recorded.

4.2 Good and Poor Ways to Sample

Biased Samples: BAD   - Systematically favor certain outcomes
- Volunteer sample-
- Convenience sample-

Random Samples: GOOD, subjects chosen by chance

Simple Random Sample:
- every set of n individuals has the same chance of being selected
- simplest type of probability sample, like taking names out of a hat
- usually done with a computer: make a list of all elements in the population (**sampling frame**) and number them, then use computer to select numbers at random
- all the statistical inference procedures we use in this class require a SRS

Sample Surveys
- Personal Interview-

- Telephone Interview-

- Questionnaires-

How Accurate Are the Results?

When using a random sample for a survey, our results will have a **margin of error** of approximately $1/\sqrt{n}$.

(We'll see how to find the exact margin of error in the next few chapters.)

Example: A poll of 1000 randomly selected registered voters is conducted a week before a major election. The results of the poll show that 51% of the sample is planning on voting for your candidate.

a) Barring any major incidents that could change people's minds, can you be confident your candidate will win?

b)   What if the poll had 55% supporting your candidate?

Other Sources of Potential Bias in Sample Surveys:
The margin of error is computed assuming there is no bias in the sampling method. It only takes into account the fact that any random sample will give results that are not identical to those in the population. In surveys in particular, you should be aware of these other potential pitfalls:
- undercoverage-

- nonresponse bias-

- response bias-

- wording of questions-

Examples:
1. A highly conservative website asks its viewers, "Do you support gay marriage?" Two thousand out of two thousand twenty respondents said that they do NOT support gay marriage. Do you believe that 99% of Americans do not support gay marriage?

2. Which of the following questions would the best for a questionnaire? Explain.
   a) Do you support an increase in tuition because it would lead to smaller class sizes and better school equipment?
   b) Do you oppose a raise in tuition because it would cause too much of a burden for students and their families?
   c) Do you support or oppose a 5% tuition increase?

3. Five hundred people were randomly selected from the phone book and asked about current apartment rental policies in town. Is this a biased or unbiased sampling method?

4. You want to conduct a survey concerning students' sexual habits, including questions such as number of partners had, age when you first had sex, etc. How could each of the following influence student's responses?
   - the interviewer's age-

   - the interviewer's gender-

   - the location of the interview-

4.3 Good and Poor Ways to Experiment

A study is an experiment when we actually do something to people, animals, or objects in order to observe the response. Here is the basic vocabulary:

Response variable- variable that we measure, so we can draw conclusions about it
Experimental units- individuals (subjects) involved in the experiment
Treatments- experimental conditions given to the subjects

Control of Variability
- **To avoid lurking variables and confounding effects,** make sure the conditions are as similar as possible for all variables except the factors being studied. In a laboratory setting we can control everything, but it is not very realistic.
- **Comparative experiment -** compare two or more groups to eliminate confounding
- **Placebo -** dummy treatment. **Placebo effect -** people really get better with a dummy treatment. Psychological effects are important when dealing with people.
- **Control group -** is the group that receives the placebo. It helps determine the true effect of the treatment. A control group is not necessary when comparing more than one treatment.

Blinding the study
- **Blind study –**
- **Double blind –**

Randomization

> **Random samples** – use a mechanical method to select subjects and assign them to treatments.
> **Advantages** of using random samples:
> - **Can use probability** to analyze the results based on random samples.
> - **Avoids selection bias:**
>   Picking fish from a common tank –
>
>   Doctors assigning patients to treatments –

Replication

- Number of **replications** = number of experimental units that get each treatment.
- Each experimental unit will react differently to the treatments. Use as many experimental units as you can, to reduce the chance of any treatment getting "lucky" - the effects of any particular experimental unit will be averaged out by the rest.

Example: Three people with headaches are given Advil, Tylenol, or Excedrin. Time until they report the pain is gone is recorded for each person.

a) Identify the following:

   response variable:

   experimental units

   treatments:

b) Control:

 c) Randomization:

d) Replication:

Statistically Significant Differences:

- If my **samples** show differences between the treatments, does that mean that there will also be differences between treatments in the **population**?
- Chance variation will almost always be present.
- Differences between two (or more) treatments are called statistically significant if they are too large to be attributed to chance.
- When a study reports a statistically significant result, it means that it found good evidence to support their hypothesis.
- **How large a difference is considered statistically significant?**   That's what we will learn in the next few chapters.

4.4 Other Ways to Conduct Experimental and Nonexperimental Studies

Multifactor Experiments

Categorical explanatory variables in an experiment are called **factors**. Some experiments have more than one factor, each with several levels. The treatments are combinations of factor levels.

Example: Nine people with headaches and nine people with muscular pain are given Advil, Tylenol, or Excedrin. Time until they report the pain is gone is recorded for each person.

- response variable:
- experimental units:
- factor(s):
- levels:
- treatments:

- replications:

Example: How do salaries for full professors compare across different races and gender? Salaries for 150 full professors at a university were determined, and the gender and race (White, Black, Hispanic, Other) for each one was recorded.

- response variable:
- experimental units:
- factor(s):
- levels:
- treatments:

- replications:

Matched Pairs Design

A more advanced form of control of variability used when there are two treatment groups.

- Each experimental unit is **matched** with another one on every possible confounder you can think of (age, height, weight, race, gender, etc.). One person from each pair gets randomly assigned to one treatment, and they are compared only against each other.
- Each person can serve as their own "perfect match", and receive the two treatments in random order. This is called a **cross-over design**.
- Some studies use identical twins in matched pairs experiments.

Blocked Design - the same idea as matched pairs, extended to three or more treatments. Each set of matched experimental units is then called a **block**.

Example: The sport of track and field has two seasons: indoor and outdoor. Does running on an indoor track versus an outdoor sand track affect runner's exertion? Ten runners run a 1500 meter distance on both an indoor track and an outdoor sand track. For both runs, the max heartbeat is recorded.

- response variable:
- experimental units:
- factor(s):
- levels:
- treatments:
- replications:

Types of Observational Studies:

- **Cross-Sectional Studies** – sample surveys that just want to take a snapshot of the population at the current time.

- **Case-Control Studies** – are retrospective studies (backward looking), in which we match each case (positive outcome) with a control (negative outcome) and then ask questions about the explanatory variable.

- **Prospective Studies** – are forward looking, and follow subjects into the future.

Other types of random samplings, useful in sample surveys but beyond the scope of this class, are cluster sampling and stratified sampling.

Chapter 5 Probability in Our Daily Lives

5.1 How Probability Quantifies Randomness

Random phenomenon- We cannot predict the next outcome, however, after many outcomes a distinct predictable pattern appears. This pattern follows the laws of probability. Random does NOT mean haphazard – it is a kind of order that emerges only in the long run.

Example: Studying the concept of probability when tossing a fair coin.

a)   What is the probability of getting a head if we toss a fair coin?

b)   Flip the coin once.
     What is the result?
     Proportion of heads obtained in all your tosses so far:
     Is this close to the probability of heads you expected? Why?

c)   Flip the coin nine more times.
     What are the results?
     Proportion of heads obtained in all your tosses so far:
     Is this close to the probability of heads you expected? Why?

d)   Imagine flipping the coin a thousand times.
     What will the results be?
     Should the proportion of heads obtained in all your tosses be closer to the probability of heads you expected? Why?

Probability Quantifies Long-Run Randomness

Probability of an outcome of a random phenomenon is the proportion of times the outcome would occur in a very long series of independent trials. Probability is a long-run **relative frequency**. **Computer simulations** are useful for "pretending" we are performing an experiment a very large number of times.

Independent trials - the outcome of any one trial is not affected by the outcome of any other.

Why do we need probability?
Probability is used in many areas: gaming theory, queuing theory, investments, weather patterns, etc.. This is a statistics course, so we will study some simple examples of probability that will help us understand the sampling distributions of statistics and the statements we make in statistical inference.

Subjective probabilities are sometimes necessary, when there is no "long series" of independent trials. They are, however, determined using all the information available.

5.2 Finding Probabilities

Sample space - the set of all possible outcomes.
Event - an outcome or group of outcomes, a subset of the sample space.
Probabilities are between 0 and 1. The sum of all outcome probabilities must equal 1.

Examples: Write down the sample space and the probability of each event.

a) Toss a fair coin and record the number of heads.

b) Toss two coins and record number of heads.

Basic Rules for Finding Probabilities about a Pair of Events
* **Complement Rule:** The probability that an event does not happen is 1 minus the probability that it does happen. For an event A, the rest of the sample space that is not in A is the complement, A^c. Then $P(A^c) = 1 - P(A)$.
* **Disjoint Events:** If two events A and B have no outcomes in common they are called disjoint events. Then the probability that one or the other occurs is the sum of their individual probabilities. That is $P(A \text{ or } B) = P(A) + P(B)$.
* **Intersection** of two events A and B is the overlap of the two events that are in both A and B.
  For disjoint events $P(A \text{ and } B) = P(A \cap B) = 0$.
* **Union** of two events A and B, consists of all events that are in A or B or both.
  In general, $P(A \text{ or } B) = P(A) + P(B) - P(A \cap B)$.
* **Independence:** If two events are independent, knowledge about one event tells us nothing about the other event. This is an informal definition, but in statistics we usually assume independence from the way sampling was conducted, we do not check it mathematically.
* **Multiplication Rule:** If A and B are independent, then $P(A \text{ and } B) = P(A) \times P(B)$.

	A	B	C	D	F
	15%	35%	30%	16%	?

Example: Grades on a very large statistics class are given according to the following distribution:

   a) What is the probability that a student gets an F in the class?
   b) What is the probability that a student gets better than a C?
   c) What is the probability that a student gets a C or better?
   d) What is the probability that two randomly selected students (independent of each other) both get an A in the class?

Example: Deal several cards from a standard deck (13 each hearts, diamonds, spades, clubs). Are all cards independent?

5.3 Conditional Probability: Probability of A Given B

Formula: Conditional Probability: $P(A \mid B) = \dfrac{P(A \cap B)}{P(B)}$

Example: Do male college students follow their school's teams more closely than females? The following data was collected in class on a Monday morning, after a particularly exciting and important basketball game. The question asked was: Did you watch the game on TV last night?

	Whole Game	Part of the Game	None of It	Total
Male	10	12	4	26
Female	21	24	30	75
Total	31	36	34	101

a) Find the conditional probability of a student having watched none of the game, given that the student was a female.

b) Find the conditional probability of a student being female, given that the student watched none of the game.

Independent Events and Conditional Probability
Recall two events A and B are independent if knowledge about one event tells us nothing about the other event. With this definition of conditional probability we can write:
- A and B are independent if: $P(A \mid B) = P(A)$
- If A and B are independent then: $P(A \text{ and } B) = P(A \cap B) = P(A) \times P(B)$

5.4 Applying the Probability Rules

Example: Random drug testing of air traffic controllers. Suppose the probability that an air traffic controller uses illegal drugs is 7/1000 (prevalence =.007). No test is perfect, but the drug test used here correctly identifies drug users with probability of 0.96 (sensitivity of the test). It also correctly identifies non-drug users with a probability of 0.93 (specificity of the test). (Agresti/Franklin)

a) If 100 non-drug users are tested, find the probability that at least one of them gets a (mistakenly) positive result.

b) Find the probability of a test giving a positive result.

c) Find the probability that a person uses drugs given that their test result was positive.

Chapter 6 Probability Distributions

6.1 Summarizing Possible Outcomes and Their Probabilities

Random Variable - is a **numerical** measurement of the outcome of a random phenomenon. We use letters such as X or Y to denote a RV.

Examples:
- Toss a coin 10 times, record number of heads.
- Roll two dice, record the sum.
- Ask 100 people if they approve of the president, record the number of "yes" answers.
- Count number of defective items in a production line.
- Record scores on an exam for each student.
- Measure the height, weight, age, or GPA of college students.

Probability Distributions for Discrete Random Variables
- A **discrete** RV has a finite list of possible outcomes.
- The **probability distribution** of a discrete random variable specifies all the values of X and their probabilities.
- **The probabilities must satisfy the following two properties**:
  - Every probability P(x) falls between 0 and 1.
  - All the probabilities add up to one.
- We can make a graph (**probability histogram**) that shows the probability distribution of a discrete random variable.

Example: Grades on a very large statistics class are given according to the following distribution:

A	B	C	D	F
15%	35%	30%	16%	4%

a) Define the random variable X as the number of grade points given for each grade. Write the probability distribution of X.

b) Is this a legitimate probability distribution?

c) What is the probability that X is less than 3?

d) What is the probability that X is less than or equal to 3?

The Mean of a Probability Distribution

Parameters – numerical summaries of populations.
μ - population mean
σ - population standard deviation

Example: How can we find the mean (average) grade given in that large statistics class?
a) Pretend there were exactly 100 students. How many should get each grade?

b) Compute the average grade for those 100 students.

c) This type of computation would work for any number of students, since the number of grades you add in the numerator must match the number of students in the denominator. In fact, you don't even have to pretend to have any number of students. Check that you get the same answer if you use the following formula:
$\mu = \sum xP(x)$ The mean of a population is also called its expected value.

Probability Distribution for a Continuous Random Variable

Continuous random variable is one that has an infinite number of possible outcomes. Its possible values can form an interval.

Smooth curve - Imagine a smoothed out histogram, representing an infinite number of people in the population.

Probability Distribution of a Continuous Random Variable
• Probabilities can be represented as areas under a density curve.
• Total area under the density curve is exactly equal to 1.
• Each outcome has a zero probability of happening, but we can assign probabilities to intervals.

Example: Random number generator – any number X in the interval between 0 and 1 is equally likely (not restricted to integers only).

1. How many sections should we divide the area into if we are interested in increments of 0.1?

2. Find the probability that X:

a) is between 0 and 0.8.

b) is between 0.2 and 0.5.

c) is exactly equal to 0.5.

d) rounds to 0.5?

e) is between 0.35 and 0.37?

6.2 Probabilities for Bell-Shaped Distributions

The Normal Distribution: A Probability Distribution with Bell-Shaped Curves
The normal distributions are a family of distributions, all of them:
- bell-shaped
- symmetric
- centered at their mean μ
- with spread given their standard deviation σ,
   which represents the change of curvature points
- notation: X ~ N(μ, σ)

Empirical Rule (Chapter 2) says that in any bell-shaped and symmetric distribution you will find approximately:
- 68% of the observations within one standard deviation of the mean
- 95% of the observations within two standard deviations of the mean
- 99.7% of the observations within three standard deviations of the mean

Standardized Observations, or z-scores
- $z = \frac{x - \mu}{\sigma}$
- The z-score tells you how many standard deviations above or below the mean an observation x is. Positive z-scores indicate values above the mean; negative z-scores indicate values below the mean.
- The distribution of z is normal with mean of zero and standard deviation of 1. This is called the **standard normal distribution**, and can be expressed as Z ~ N(0, 1).

Example: The distribution of IQ scores is approximately normal with mean 100 and standard deviation 16. We can write this as X ~ N(100,16).

a) If John has an IQ of 125, what is his z-score?

b) If Jim has an IQ of 87, what is his z-score?

Using the Standard Normal Table
- The standard normal table appears in the appendix of your book.
- It gives areas under the normal curve to the left of z – these are called **cumulative probabilities**.
- The z-scores appear on the margins of the table, areas are in the center.

Finding Normal Probabilities:
- Determine what the problem is asking for in terms of x.
- Standardize x by using the z-score.
- Draw the Standard Normal Curve and shade the area asked for in the problem.
- Determine the area by using the standard normal table (z table).

Example: Refer to the distribution of IQs above. What percentage of people have IQs:

a) lower than 123?

b) lower than 89?

c) higher than 89?

d) between 89 and 123?

e) exactly 89?

Finding the Value of x Given a Proportion.
- Draw a picture, with the area given shaded on it.
- Look up the cumulative area in the middle of the z table, and look at the margins to find the z-score corresponding to that area.
- Use the z-score formula to solve for x.

Example: Refer to the distribution of IQs above. What IQ corresponds to:

a) the bottom 20%?

b) the top 5%?

Example: SAT scores for each section are standardized so scores follow an approximately normal distribution with mean 500 points and standard deviation 100 points. We can write this as $X \sim N(500,100)$. The maximum score possible is 800 points.

a)   Sketch the distribution and locate the mean and change of curvature points.

b)   What proportion of students will score above 700 in one section of the SAT?

c)   What proportion of students will score below 260 in one section of the SAT?

d)   What proportion of students will score between 400 and 500 in one section of the SAT?

e)   What score corresponds to the top 10%?

f)   What score corresponds to the bottom 8%?

g)   Between what two values will you find the central 95% of scores?

h)   Between what two values will you find the central 90% of scores?

Example: You scored 650 on the SAT. Your friend took the ACT instead, and scored 30. The ACT that year had a mean of 21.0 and standard deviation of 4.7. Who did better?

Example: Camera prices have a mean of $500 with a standard deviation of $200. What can you say about the shape of the distribution of prices?

Example: A beauty product manufacturer produces and bottles body lotion for several different size containers. Suppose that they are filling the bottles according to the label with 236mL of lotion. Assume that contents in the bottles are normally distributed.

a) Do all of the bottles contain the same amount of lotion?

b) Suppose the machine is set so that, on average, bottles will contain 236mL of lotion. How many bottles will contain less than the amount specified on the label?

c) As a consumer, would you consider this fair?

d) What should be done to the mean setting so that more of the bottles contain the label amount?

e) Usually, the mean setting for the machine is easy to adjust (turn of a knob) but the variability depends on the precision of the machine. Suppose the machine is set so that the distribution of bottles contents is approximately normal, with mean of 240mL and standard deviation of 3mL. What proportion of bottles will contain less than the label says?

6.3 Probabilities When Each Observation Has Two Possible Outcomes

The Binomial Distribution: Probabilities for Counts with Binary Data

Conditions for Binomial Distribution for Count of Successes:
- There are a fixed number of trials, n.
- The n trials are all independent.
- The outcome of each observation is either a "success" or a "failure."
- Each trial has the same probability of success, p.

Then the distribution of X = number of successes is called a **binomial distribution** with parameters n and p.

Examples – Are They Binomial? If yes, identify n and p. If not, explain why not.

- A perfectly balanced coin lands heads up 50% of the time. Toss a balanced coin n times and record the number of heads.

- Toss a balanced coin until we see 10 heads. Record the number of tosses.

- To estimate the proportion of underage drinkers at a very large university, you ask 100 randomly selected students if they drank (or drink) when they were (are) underage. Make sure that the questioning is handled in a way that the respondent's answer would be confidential. Count the number of students who answer yes to the question.

- To estimate the proportion of underage drinkers in a class of 20 students, you ask 10 randomly selected students if they drank (or drink) when they were (are) underage. Make sure that the questioning is handled in a way that the respondent's answer would be confidential. Count the number of students who answer yes to the question.

Probabilities for a Binomial Distribution

$$P(x) = \binom{n}{x} p^x (1-p)^{n-x} \quad \text{for } x = 0, 1, 2, 3, \ldots, n$$

$$\text{where } \binom{n}{x} = \frac{n!}{x!(n-x)!} \quad \text{and} \quad n! = n \cdot (n-1) \cdot (n-2)\ldots 3 \cdot 2 \cdot 1$$

Mean: $\mu = np$

Standard Deviation: $\sigma = \sqrt{np(1-p)}$

Example: Toss a balanced coin three times and count the number of heads.

a) Write all of the possible outcomes for three coin tosses.

b) Write the sample space.

c) Write the probability distribution of X.

d) Find the number of ways that you can get two heads when you flip the coin three times using the formula above and compare to the results from part a.

e) Use the Binomial formula to find the probability of exactly two heads.

f) Find the mean of this situation.

g) Find the standard deviation.

Example: A particular medication causes side effects on 35% of patients. Eight patients at our clinic are currently receiving that medication.   Let X= number of these patients that experience side effects.

a)   Is the distribution of X binomial? Explain what assumptions we need to make.

b)   Write down the sample space for this distribution.

c)   How many patients out of 8 would you expect to experience side effects?

d)   Find the standard deviation of this distribution.

e)   Find the probability that 6 out of the 8 patients experience side effects.

f)   Find the probability that 3 out of the 8 patients experience side effects.

g)   Find the probability that at least one patient gets side effects.

Chapter 7 Sampling Distributions

Commonly, we only think of taking one sample. What if we took multiple samples? How does the first sample's results compare to the population and to other samples?

We are going to look at two different situations.
- Sample Mean
    Data:
- Sample Proportion
    Data:

7.1 How Sample Proportions Vary Around the Population Proportion

Example: Suppose 30% of the population smokes (p=.30) and that we took a sample of 100 randomly selected individuals from the population and 27 said that they smoked.
a) What would be the population distribution?

b) What would be the sample distribution (also known as the data distribution)?

c) What would be X for this sample? _____
d) What would be $\hat{p}$ for this sample? _____

e) What does $\hat{p}$ estimate? _____
f) What would be the results if we took a different sample?

g) What would be possible values of X?

h) What would be possible values of $\hat{p}$?

i) What would be the results if each student in the class took their own sample and reported the results to the class?

The Sampling Distribution of a Statistic:

Ideas Behind the Sampling Distribution:
- Statistics are random variables.

- Random variables have a distribution.

- The distribution of a statistic is called a _____ .

- How can we study the sampling distribution of a statistic?

Sampling Distribution of $\hat{p}$

We want to know the distribution of all the possible values the sample proportion $\hat{p}$ can take in repeated sampling.

- **Does** $\hat{p}$ **have a binomial distribution?**

- **Normal approximation**:

- **Mean of the distribution of** $\hat{p}$ **:**

- **Standard error of the distribution of** $\hat{p}$ **:**

    Note – we use the term **standard error** to refer to the standard deviation of a sampling distribution, and to distinguish it from the standard deviation of an ordinary probability distribution.

Sampling Distribution of $\hat{p}$

is approximately normal, with mean = p and standard error = $\sqrt{\dfrac{p(1-p)}{n}}$

as long as the expected number of successes (np) and failures (n(1-p)) are each 15 or larger.

Example: Suppose that 54% of the registered voters in the population support your favorite candidate – that is, if the election were today, your candidate will win. A survey before the election asks a random sample of 1000 registered voters whom they are planning to vote for in the election. Find the probability that the survey concludes your candidate will win – that is, find the probability that more than 50% of voters in the sample say they will vote for your candidate.

a)   Can we use the normal distribution to approximate probabilities?

b)   What is the sampling distribution of the sample proportion?

c)   Approximate the probability that the proportion of successes in a sample is more than 50% using the normal table.

Example: A sample of 20 students is asked if they prefer Coke or Pepsi. Suppose that in the population, 80% prefer Coke. Find the probability that more than 2/3 of students in the sample prefer Coke.

a)   Can we use the normal distribution to approximate probabilities?

b)   What if we ask 200 students?

7.2 How Sample Means Vary Around the Population Mean

Example: Consider the distribution of heights of 20 year old women. Suppose the mean is 65" and the standard deviation is 2.5". A sample of 5 women was taken and their heights were 60, 62, 64.5, 66.5, and 67 inches.

a)   What would be the population distribution?

b)   What would be the sample distribution (also known as the data distribution)?

c)   What would be X for this sample? _____

d)   What would be $\bar{x}$ for this sample? _____

e)   What does $\bar{x}$ estimate? _____

f)   What would be the results if we took a different sample?

g)   What would be the results if each student in the class took their own sample and reported the results to the class?

Example: Consider the distribution of heights of 20 year old women. Suppose the mean is 65" and the standard deviation is 2.5". What do you expect will be the shape of this distribution?

a) Sketch this distribution. Use a • to represent the **height of each woman**.

b) Now consider the distribution of the average height of 5 women at a time. Use a different symbol to represent the **average height of each sample**.

c) Now consider the distribution of the average height of 30 women at a time. Use the same symbol as above to represent the **average height of each sample**.

Example: Consider the distribution of the last digits in the serial numbers of $1 bills.

a) Sketch this distribution. Use a ● to represent the **digit in each dollar bill.**

b) Now consider the distribution of the average of 5 of those digits. Use a different symbol to represent the **average of all digits in each sample**.

c) Now consider the distribution of the average of 50 of those digits. Use the same symbol as above to represent the **average of all digits in each sample**.

What can we tell about the sampling distribution of $\bar{x}$ from the previous examples?

- Mean of the sampling distribution of $\bar{x}$

- Standard error of the sampling distribution of $\bar{x}$

- Shape of the sampling distribution of $\bar{x}$

Central Limit Theorem CLT
Sampling Distribution of the Sample Mean

For a random and representative sample (SRS) with a <u>large</u> sample size n, the
sampling distribution of the sample mean is
approximately **normal** with **mean** μ (same as the original distribution) and

standard error $\dfrac{\sigma}{\sqrt{n}}$ (the original standard deviation divided by the square root of n).

How large does the sample size, n, have to be?
- It depends on the shape of the original population.

- If the population is normal, the sampling distribution of $\bar{x}$ will be normal for any n.

- If the population is far from normal, n=30 is large enough in most cases for the sampling distribution of $\bar{x}$ to be considered normal.

- In general, the closer to normal (bell, shaped, symmetric, and continuous) the original distribution is, the smaller n needs to be.

- And for any shape distribution, as n increases, the sampling distribution of $\bar{x}$ will get closer to normal.

Examples:

	Original Population	Sampling Distribution of $\bar{x}$ for n=2	Sampling Distribution of $\bar{x}$ for n=5	Sampling Distribution of $\bar{x}$ for n=30
Uniform				
Bimodal				
Skewed Right				
Bell-shaped				

Example: Commercial food is packaged on vast assembly lines. A particular size of soup can is supposed to have 19oz, but there is variability from can to can – it is a random variable. Suppose that the distribution of contents is approximately normal with mean of 19.1 oz. and standard deviation 0.05.
a) Find the probability that one randomly selected can contains less than 19oz.

b) Wholesale warehouses sell these cans in packs of four. Find the probability that the average of a randomly selected package of four is less than 19oz.

c) Between what two values would you find the average contents of central 95% of four-packs?

d) Why could we use the normal table to find the probabilities in this problem?

Working on Sampling Distribution Problems

Sample Proportion:
- Data: _____

- What is X? _____

- What is $\hat{p}$? _____

- What is the sampling distribution of the statistic?

- What are the assumptions?

Sample Mean

- Data: _____

- What is X? _____

- What is $\bar{x}$? _____
- What is the sampling distribution of the statistic?

- What are the assumptions?

Examples: Identify the type of problem (sample mean or sample proportion) and determine if the sampling distribution of the statistic is approximately normal before computing the probabilities.

1. Suppose that 40% of residents in a city support construction of a new Wal-Mart supercenter. A sample of 400 adults was taken. What is the probability that at least a half of the sample supports the supercenter?

    a)   Is this problem about the sampling distribution of the sample proportion or the sample mean?

    b)   What is the sampling distribution of the statistic?

    c)   Can we answer the question posed? If so, answer the question, if not, explain why not.

2. The distribution of lawyer's salaries at a firm has mean of $90,000 and standard deviation of $50,000. What is the probability that the average salary of a random sample of five lawyers from this firm is less than $100,000?

    a)   Is this problem about the sampling distribution of the sample proportion or the sample mean?

    b)   What is the sampling distribution of the statistic?

    c)   Can we answer the question posed? If so, answer the question, if not, explain why not.

3. The distribution of lawyer's salaries at a firm has mean of $90,000 and standard deviation of $50,000. What is the probability that the average salary of a random sample of forty lawyers from this firm is less than $100,000?

   a)   Is this problem about the sampling distribution of the sample proportion or the sample mean?

   b)   What is the sampling distribution of the statistic?

   c)   Can we answer the question posed? If so, answer the question, if not, explain why not.

Example: Suppose that X= number of heart attack patients arriving at a clinic per week. The distribution of X has a mean of 3.5 patients, with a standard deviation of 1.7.

   a)   Can the random variable X have a binomial distribution?

   b)   Can the random variable X have a normal distribution?

   c)   Suppose that we were interested in $\bar{x}$ equal to the average number of heart attacks per week in one year. What would n be? _____

   d)   Find the sampling distribution of the sample mean.

   e)   Find the probability that, in one year, there is an average of less than 3 heart attacks per week at the clinic.

How Far Off Will a Sample Prediction Be?

For both means and proportions, we saw that the normal distribution can be used to approximate the sampling distribution of the statistic. That is the reason the normal distribution is the most important one in statistics. Since normal distributions follow the Empirical Rule, we know that:

- 99.7% of sample results will fall within 3 standard errors of the true parameter

- 95% of sample results will fall within 2 standard errors of the true parameter

Chapter 8 Statistical Inference: Confidence Intervals

8.1 Point and Interval Estimates of Population Parameters

Point Estimation
We use statistics to estimate parameters. Fill in the chart below with the statistic that estimates each of the following parameters.

Symbols:	Population Parameter	Sample Statistic
mean		
standard deviation		
proportion		

Parameters vs. Statistics
- If we were to take a sample and calculate the sample mean, would the sample mean be exactly equal to the population mean?

- If we were to take a sample and compute the sample proportion, would the sample proportion be exactly equal to the population proportion?

- These statistics are called **point estimates**. They are our "best guess" of the parameter, but we need to know more about their variability and potential biases.

Bias and Standard Error – all the statistics we use in this class are unbiased and have small variability.

Bias has to do with the **center** of the sampling distribution.
- An unbiased statistic has its sampling distribution centered at the parameter being estimated.
- How do you reduce bias?

Standard error has to do with the **spread** of the sampling distribution.
- A smaller spread means that we have more values of the estimate closer to the parameter being estimated.
- How do you reduce standard error?

Example: Throwing darts.

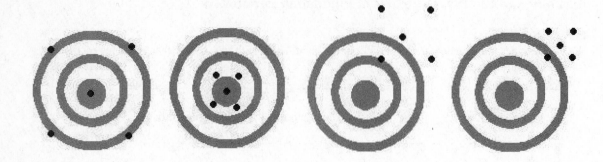

8.2 Constructing a Confidence Interval to Estimate a Population Proportion

Example: Suppose we want to know the true population proportion (p) of Americans who are willing to accept a cut in living standards to help the environment. In 2010, the General Social Survey asked a random sample of Americans how willing they were to accept a cut in living standards to help the environment. Out of 1,381 respondents, 496 said that they were very willing or fairly willing. What can we say about p?

a)   What is the sample proportion?

b)   Is p equal to the sample proportion?

c)   What is the sampling distribution of $\hat{p}$?

d)   Where will we find the central 95% of all the sample proportions?

So 95% of all samples will be "lucky" enough that their $\hat{p}$ will be within 1.96 standard errors of p. We can flip the reasoning around and say that p will be within 1.96 standard errors of $\hat{p}$ for 95% of all samples.

e)   The exact standard error of the sample proportion equals: $\sqrt{\dfrac{p(1-p)}{n}}$.

Do we know p?
What can we use to estimate p in the formula for the standard error?

f)   95% confidence interval for p:

g)   Compute the 95% confidence interval for the example and interpret.

h)   Is p equal to 0.34?

i)   Is p equal to 0.38?

j)   Is p equal to 0.30?

k)   Does our interval include the true population proportion of all Americans that are willing to accept a cut in standard of living to help the environment? Is our sample one of the "lucky" ones?

We will never know. But probability is in our favor. For 95% of all random samples, that formula will produce an interval that contains the unknown parameter p and only 5% of samples will give intervals that miss p.

For a simulation of what would happen if we took a random sample, constructed a CI, and then repeated this process over and over again, check out the Confidence Interval applet called "Confidence Intervals for the Proportion."

Confidence intervals are of the form: estimator ± margin of error

- In this case, the estimator of p is $\hat{p}$.
- The margin of error depends on two things: the confidence level we want, and the standard error of our estimator.
- For any level of confidence that you want, you can find a value, z, in the z table, that will tell you how many standard deviations you need to go left and right of $\hat{p}$.
- Typically, confidence levels are at least 90% -- most common are 90%, 95%, and 99%.
- Complete the table below.

Confidence Level	Tail Area	z
90%	.05	1.645
95%	.025	1.96
99%	.005	2.58
92%		
97%		

Confidence Interval for a Population Proportion

For a random and representative sample (SRS), the confidence interval for p is:

$$\hat{p} \pm z\sqrt{\frac{\hat{p}(1-\hat{p})}{n}}$$

This interval is only valid when you have at least
15 successes ($n\hat{p} \geq 15$) and 15 failures ($n(1-\hat{p}) \geq 15$).

Example: Find the 99% confidence interval for the proportion of Americans who are willing to accept a cut in living standards to help the environment

8.3 Constructing a Confidence Interval to Estimate a Population Mean

Confidence intervals for population means are similar in form to those made for a population proportion.:
estimator ± margin of error

Confidence interval for the population mean, μ
- **Estimator** of μ is $\bar{x}$.

- **Margin of error**, again, depends on the confidence level we want, and the standard error of our estimator.

- What would be your guess of the confidence interval for the population mean?

- Unfortunately, when we don't know μ, we also don't know σ. What can we use to estimate it?

- Because we are estimating σ with s, we have to use a different distribution, instead of the z table. We use the t table with n-1 degrees of freedom (associated with s).

- Formula for confidence interval for the population mean, μ

The t distributions:
- family of distributions indexed by their degrees of freedom (df)
- all symmetric and bell-shaped, all centered at zero
- more spread out than z (fatter tails, lower peak)
- as df increase, t gets closer to z
- smaller sample ⟹ more variability ⟹ fatter tails

Finding Probabilities Using the t Table
- Table B, in the appendix of your book, gives t scores for the t distribution.
- It is read very differently from the z table: df along the left margin, right tail areas and confidence levels appear along the top, and values of t in the middle of the table.
- The notation for a t distribution is $t_{right\_tail\_area}$, in other words, a $t_{.01}$ is a t-score with probability 0.01 to the right.

Examples: Find the t-score for the following conditions:

1. t-score that has probability .01 to the right of it in a t distribution with 12 degrees of freedom.

2. t-score that has probability .95 to the left of it in a t distribution with 30 degrees of freedom.

3. t-score that has the probability 0.01 to the left in a t distribution with 13 degrees of freedom.

4. t-score that would be used in a 95% CI for a sample of n=15.

5. t-score that would be used in a 95% CI for a sample of n=100.

Confidence Interval for a Population Mean

For a random and representative (SRS) sample, the confidence interval for μ is:

$$\bar{x} \pm t \frac{s}{\sqrt{n}}$$

The t-score for the CI comes from a t distribution with n-1 degrees of freedom.
This is valid when the original distribution is normal or n is large.

Small sample size:
- t procedures are very sensitive to skewness or outliers in the original population
- s might be far from σ
- t distribution still far from normal
- **need to use t table AND need original population to be normal**
- impossible to check population, we only have a small sample
- plot data and make sure it **could have come** from a normal distribution. Perfect symmetry of the sample not important, but there should be **no major outliers**

Large sample size:
- CLT guarantees that the sample mean has a normal distribution when n is large
- s is a good estimator of σ
- t distribution gets close to normal -- think of the z distribution as a t with df = infinity
- makes very little difference to **use z or t**
- if df are not on table we can use the z table instead. Minitab can give us exact values.

SRS: Regardless of sample size, the data must be a simple random sample from the population of interest, in order to extend the conclusions.

Example: High efficiency washing machines use less water and less detergent than standard washing machines. Everyone varies in the amount of detergent that they put into a washing machine. Some people might put in a little bit more; while others might use a little bit less. How many loads does a full 150 ounce bottle of leading brand laundry detergent handle on average? The data is below.

$$87 \quad 88 \quad 103 \quad 98 \quad 94 \quad 99 \quad 89 \quad 91 \quad 87 \quad 91$$

a) Are the assumptions met to make a 95% confidence interval for the population mean?

b) Make a 95% CI for μ and interpret.

c) The Minitab output is below for this problem. Does it agree with what you have done?

```
One-Sample T: detergent

Test of mu = 96 vs not = 96

Variable    N    Mean   StDev   SE Mean       95% CI           T      P
detergent   10   92.70   5.60      1.77   (88.69, 96.71)   -1.86  0.095
```

Example: An investor was interested in determining the average monthly change in her investment portfolio. She randomly selected 7 monthly statements and recorded the change in the investment for each of the months. These are listed below.

$487.64 $1106.30   -$92.66 $463.27 $273.96 $572.73 $196.38

a)   What does the negative sign in front of 92.66 mean?

b)   Below is a boxplot of the data above. Is it reasonable to use the t procedures here?

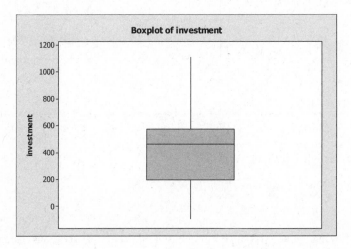

c)   Below is the Minitab output for a 99% confidence interval. Interpret.

One-Sample T: Investment

```
Variable   N     Mean    StDev    SE Mean        99% CI
Investme   7    429.660  372.890  140.939   (-92.862, 952.182)
```

How Confidence Intervals Behave: $\bar{x} \pm t \dfrac{s}{\sqrt{n}}$

- These CIs for μ are always centered around $\bar{x}$.
- The length of the interval depends on 3 things:
  - the confidence level (determined by t)
  - the standard deviation, s
  - the sample size, n
- Typically, we have no control over s, but we can control the confidence level and the sample size, n.

Example: Using the data about the number of loads of laundry, we found that the sample mean was 92.7 and the sample standard deviation was 5.60.

a) Compute the following CIs for μ:

Conf. Level	$\bar{x} \pm t \dfrac{s}{\sqrt{n}}$	CI
99%	92.7 ± (5.60 / $\sqrt{10}$)	
95%	92.7 ± (5.60 / $\sqrt{10}$)	
90%	92.7 ± (5.60 / $\sqrt{10}$)	

b) Mark these three CIs on the number line.

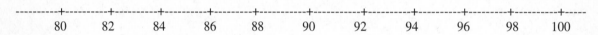

c) How does the confidence level affect the length of the interval?

d) How would the 95% interval change if we had gotten the same $\bar{x}$, but based on twenty observations instead of eight?

e) Mark these two CIs on the number line.

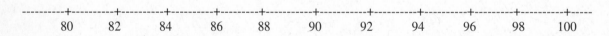

In general:
- Bigger Confidence ⇒ Bigger Interval
- Bigger Sample Size ⇒ Smaller Interval
- If you want to have a lot of confidence, but not a huge interval, increase the sample size.

Important Notes about Interpreting Confidence Intervals
- Confidence intervals are statements about the POPULATION MEAN, not about the sample mean or about individuals.

- We can only talk about "probability" BEFORE we take the sample. After that, we talk about "confidence." Another branch of statistics called Bayesian statistics does not make this distinction.

Examples: Suppose the 95% confidence interval for the average number of hours students study for each class in one week is (2.3, 5.4). Determine what is wrong (if anything) with each interpretation.

1. We are 95% confident that the true average number of hours students spend studying for one course in one week is between 2.3 and 5.4 hours.

2. We are 95% confident that all students study between 2.3 and 5.4 hours for one course in one week.

3. The probability that a student studies between 2.3 and 5.4 hours for one course in one week is 0.95.

4. The probability that the true mean, μ, is in the 95% confidence interval is 0.95.

5. We are 95% confident that the sample mean number of hours that students study is between 2.3 and 5.4 hours for one course in one week.

Working on Confidence Interval Problems

- **CI for Population Mean:** $\bar{x} \pm t\dfrac{s}{\sqrt{n}}$

- **CI for Population Proportion:** $\hat{p} \pm z\sqrt{\dfrac{\hat{p}(1-\hat{p})}{n}}$

Examples: For each of the following stories, identify the type of problem (mean or proportion), check the necessary assumptions, construct the confidence interval, and interpret the results.

1. A survey asks 231 college students how many sexual partners they have had in their lives. The sample mean was 4.641 and the sample standard deviation equals 6.33.

a)   Is the parameter being estimated the population proportion or the population mean?

b)   Are the assumptions for a confidence interval for the population parameter met?

c)   Construct a 95% CI for the parameter.

d)   Interpret the interval.

2. In 2010, the GSS asked participants their opinions about spanking. They asked the following question, "Do you strongly agree, agree, disagree, or strongly disagree that it is sometimes necessary to discipline a child with a good, hard spanking?" Out of 1416 respondents, 976 said that they agreed or strongly agreed that it was sometimes necessary.

a) Is the parameter being estimated the population proportion or the population mean?

b) Are the assumptions for a confidence interval for the population parameter met?

c) Construct a 99% CI for the parameter.

d) Interpret the interval.

3. As part of a small initial case study, sixteen patients were randomly selected from a hospital's record of chemotherapy patients. They were asked to participate in a trial of a new type of drug to ease queasiness during chemotherapy based on ginger. After receiving the new queasiness drug, the patients were asked to record if they felt reduced queasiness or not reduced queasiness from their next dose of chemotherapy. Ten out of the sixteen patients said they had reduced queasiness with the new medication.

a) Is the parameter being estimated the population proportion or the population mean?

b) Are the assumptions for a confidence interval for the population parameter met?

c) Construct a 95% CI for the parameter.

d) Interpret the interval.

e) Why is the width of the interval so wide?

f) If we wanted a smaller interval but still with 95% confidence, what could we do?

8.4 Choosing the Sample Size of a Study

When planning a study, we can determine how large the sample needs to be to estimate the parameter within a given margin of error, with the desired confidence.

Determining Sample Size for Estimating a Population Proportion

Margin of error for a CI for p is: $m = z\sqrt{\dfrac{\hat{p}(1-\hat{p})}{n}}$

This margin of error depends on $\hat{p}$, which we won't know until we conduct the actual study (remember -- we are doing this in the planning stage of the study). We need another way to guess what $\hat{p}$ will be.

Choosing the right educated guess for p in the standard error:

The standard error is largest when $\hat{p} = 0.5$. That means that the sample size needed for a particular margin of error will be largest when $\hat{p} = 0.5$. The value of n won't change much if $\hat{p}$ is close to 0.5, but it will be considerably smaller if $\hat{p}$ is close to 0 or 1. A good guess can dramatically decrease the sample size needed.

- Use $\hat{p} = 0.5$ if you have no clue what the true value of $\hat{p}$ is.
- Use a guess for $\hat{p}$ from a previous study, if available.

Solving for n:
- Decide on a value for $\hat{p}$ as your educated guess.
- Use the confidence level and margin of error given in the problem.
- The minimum sample size needed for achieving the desired margin of error is found by solving the above equation for n:

- Answer must be an integer. If not, **round up** to the next largest integer.

Example: You want to know what proportion of students is financially supported by their parents. Find the number of students that should be sampled to have a margin of error equal to 0.02, with 95% confidence.
a) If you have no clue about the proportion:

b) If a previous study last year found that the proportion was near 0.90:

Determining Sample Size for Estimating a Population Mean

Margin of error for a CI for μ is: $m = t\dfrac{s}{\sqrt{n}}$

The value of t depends on the degrees of freedom of the problem (n-1). Since we don't know n, we cannot determine the value of t. However, we do know that with large sample sizes, the values of t approach the normal distribution.

Solving for n:

- Use the confidence level, margin of error, and standard deviation estimate given in the problem.
- The minimum sample size needed for achieving the desired margin of error is found by solving the above equation for n:

- Answer must be an integer. If not, **round up** to the next largest integer.

Example: We are trying to determine how many 1 oz. containers of water we should collect from a potentially contaminated stream. We want our estimate of the average pH of the stream to be accurate to within ± 0.03, with 95% confidence. We know from a preliminary study that the standard deviation is around 0.25. How many 1 oz. containers should we collect?

Example: We would like to estimate the true average GPA of all students to within ± 0.05 points, with 99% confidence. We know that for this school the GPA ranges from 0 to 4. How many students should be sampled?

Chapter 9 Statistical Inference: Significance Tests about Hypothesis

9.1 Steps for Performing a Significance Test

There are two different types of statistical inference:

Confidence Intervals (previous chapter):
- Give a region that is likely to contain the parameter.
- We have no preconceived notion of what our parameter should be; we simply want to estimate it.

Significance Tests (this chapter):
- Check to see which claim about the population is supported by the data.
- Someone proposes a value of a parameter. We disagree with that value, so we take a sample to try to see if the data supports that claim, or if it supports what we believe is true.
- They have a very elaborate vocabulary, but the basic idea behind them is quite simple.

The Ideas Behind the Tests of Significance

Example: Is our favorite candidate going to win the next election for mayor? If more than 50% vote for your candidate, your candidate will win. If 50% or less of the voters picks your candidate, there will be a runoff between the top two candidates. You decide to take a simple random sample of registered voters in our town. Out of the 200 registered voters in your sample, 114 said that they would vote for our candidate. Is this evidence that the majority of registered voters in your town support our candidate?

Solution: In order to win, our candidate needs more than 50% of the vote. Pretend that the proportion of people who will vote for our candidate is 0.5 – there is a runoff, and determine how likely (or unlikely) it would be to get results as high as those in the sample.

a) If we pretend that the proportion of people in the town who would vote for our candidate is equal to 0.50, what is the distribution of the sample proportions from all possible samples?

b) According to that distribution, what is the probability that a random sample would give a result as high or higher than the one you observed?

c) Considering that this is a very unlikely result, what can you conclude about the claim that the proportion that the population proportion of people who are going to vote for our candidate is equal to 0.50?

Elements of a Significance Test	In Our Example:
Assumptions	
Null Hypothesis: H$_o$ **Alternative Hypothesis: H$_a$**	
Test Statistic: z-score summarizes the information from the sample - measures how far away the point estimate is from the value of the parameter specified in the null hypothesis in terms of standard errors	
p-value: "corner" area probability that the test statistic equals the observed value or a value even more extreme if the null hypothesis is true	
Conclusions statement based on the p-value in everyday language **Small p-values support H$_a$.**	

Stating the Hypotheses

- **Statements about the Parameters**: The null and alternative hypotheses are always statements about the unknown parameters, not the sample statistics.

- **Null Hypothesis** states that the parameter takes on a particular value
  $$H_o: p = \#$$

- **Alternative Hypothesis** can be either
  - **one-sided** - we are only interested in deviations from H_o in only one direction:
    $$H_a: p < \# \quad \text{or} \quad H_a: p > \#$$

  - **two-sided -** we simply want to show that the parameter is different from the number claimed in the null hypothesis:
    $$H_a: p \neq \#$$

- **Sign of the Alternative**: We always determine the sign of the alternative hypothesis from the story - not from the data.

Examples: Write down the null and alternative hypotheses for each of the following scenarios. Define p for each one.

1. In a particular town, the proportion of accidents per year that involve people talking on a cell phone is 0.40. A new ad campaign was conducted by the Highways Department to encourage people to pull over when they talk on a cell phone. Does the proportion of accidents with cell phones <u>decrease</u>?

2. A student senator needs <u>more</u> than 2/3 of the votes in order to pass a ruling that increases student activity fee spending on non-fraternity or sorority events. He takes a small sample to see if there is support for his bill. Is there evidence to show that more than 2/3 of student senators may vote for this bill?

3. The proportion of male college students that binge drink is close to 0.5. Is it <u>different</u> for women?

P-value and Statistical Significance

The p-value is a probability, and thus a number between 0 and 1. It represents the probability that the test statistic equals the observed value or a value more extreme, if in fact, H_o were true.

What does extreme mean? Extreme here means too far from the value specified in H_o -- too far in the direction specified by H_a.

What kind of test statistic will yield a small p-value?

Interpreting the p-value. The p-value represents the strength of the evidence for the null hypothesis. The smaller the p-value, the more evidence <u>against</u> the null and <u>for</u> the alternative hypothesis (the one we are trying to determine if there is evidence to support).

How small should the p-value be for there to be evidence for H_a? The smaller the better.

Example: Match the following p-values with the strength of the evidence against H_o.

Conclusions	P-values
___ 1. very strong evidence against H_o	a) 0.15
___ 2. strong evidence against H_o	b) 0.06
___ 3. some evidence against H_o	c) 0.02
___ 4. not enough evidence against H_o	d) 0.0002

Significance Level (α level). In practice, we usually determine if the p-value is "small enough" by comparing it to a pre-specified significance level, α. The usual α levels are 0.10, 0.05, and 0.01, which correspond to 90%, 95%, and 99% confidence, respectively.

P-values and Statistical Significance

Statistical Significance:
If the p-value $\leq \alpha$, we reject H_o at an α level of significance.

Interpreting p-values:
- **Small p-value $\Rightarrow$ Evidence that H_o is wrong (or we got an unlucky sample)**
  - **$\Rightarrow$ Results are significant (Evidence for Ha)**
  - **$\Rightarrow$ Reject H_o**

- **Large p-value $\Rightarrow$Not sufficient evidence to show that H_o is wrong**
  - **$\Rightarrow$ Results are NOT significant**
  - **$\Rightarrow$ Fail to Reject H_o**

Examples:
1. p-value = 0.03

2. p-value = 0.0004

3. p-value = 0.12

9.2 Significance Tests about Proportions

Elements of a Significance Test for a Population Proportion:

1. **Assumptions**
   -
   -
   -

2. **Hypothesis**
   - **Null Hypothesis**, H_o

   - **Alternative Hypothesis,** H_a

3. **Test Statistic**, TS
   z-score, summarizes the data

4. **P-value:** probability that the test statistic equals the observed value or a value even more extreme if the null hypothesis is true.

5. **Conclusion:**
   Small p-value $\Rightarrow$ Reject H_o
   Large p-value $\Rightarrow$ Fail to Reject H_o (never accept H_o)

Example: Not getting enough sleep is a common complaint from college students. Are sleep problems common for American women as well? In a discussion in a health class, one of the students in the class says that only about 50% of American women have trouble sleeping. You don't think that this is correct. You are fairly certain that the student's value is too low, but you decided to be cautious and make a two-sided test to see if the proportion of American women that have trouble sleeping is different from 50%. You decide to go to the GSS survey page to research the topic. You find information from the year 2010. Out of 617 women, 362 said that they often or sometimes had trouble falling to sleep or staying asleep in the past year. Is there evidence to show that the population proportion of American women who often or sometimes have trouble sleeping or staying asleep in the past year is different from 0.5?

Example: A very large manufacturing plant offers management safety classes to each of its employees on the floor. The class is mandatory for advancement, but space is limited in the class. About 30% of the employees on the floor are women. Women claim that they are not getting equal access to the management safety class and thus to further advancement. Out of the 80 past students in the management safety class, 21 of the students were women. Is there statistically significant evidence of gender bias in enrollment in the management safety class?

Example Revisited: For the preceding example, answer the following questions

a)   Write down the null and alternative hypothesis for this test.

b)   What kind of values for $\hat{p}$ would support:

     the alternative hypothesis?

     the null hypothesis?

c)   For each of the following values of $\hat{p}$, find the test statistic, p-value, and the conclusions.

$\hat{p}$	TS	p-value	conclusion
0.15			
0.20			
0.225			
0.25			
0.35			

9.3 Significance Tests about Means

Elements of a Significance Test for a Population Mean

1. **Assumptions**
   -
   -
   -

2. **Hypothesis**
   - **Null Hypothesis**, H_o

   - **Alternative Hypothesis**, H_a

3. **Test Statistic**, TS
   t-score, summarizes the data

4. **P-value:** probability that the test statistic equals the observed value or a value even more extreme if the null hypothesis is true.

5. **Conclusion:**
   Small p-value $\Rightarrow$ Reject H_o
   Large p-value $\Rightarrow$ Fail to Reject H_o (never accept H_o)

Example: A fellow student says that the average salary of graduates in your major is 30,000 dollars per year. You don't think that this is correct. You think that the average salary is higher than this. To show that your fellow student is wrong, you take a simple random sample of fifty graduates who have graduated in the past 5 years and ask for the amount of their starting salary. You find that the sample mean is 31,000 and the sample standard deviation is 4,000. Is there evidence to support your claim that the population mean starting salary is greater than 30,000?

Example: An environmentalist is interested in determining if the pH of the creek water behind his house is affected by the new development upstream. He knows that a neutral stream has a pH of 7. He takes 16 samples of water and determines the pH. Is the average pH of the creek significantly different from neutral? Conduct a significance test at the 0.01, 0.05 and 0.10 significant levels.

7.5	7.6	7.1	6.2	6.3	6.9	7.1	7.3
6.1	6.3	6.5	7.1	6.3	6.7	6.9	6.3

Relationship between CI and Significance Test

Example: Use the data the environmentalist collected in the preceding example to construct confidence intervals and compare the results to those of the significance test.

a) Make a 95% CI for μ:

b) Compare the results of the confidence intervals to those of the significance test at $\alpha = 0.05$.

c) Make a 90% CI for μ:

d) Compare the results of the confidence intervals to those of the significance test at $\alpha = 0.10$.

Relationship between
Confidence Intervals and Two-Sided Tests:

The results of any two-sided test (H_o: $\mu = \mu_o$ vs. H_a: $\mu \neq \mu_o$) and confidence interval on the same set of data must agree:

- If the p-value of a test is less than α, a $(1-\alpha)*100$ % CI does not contain the H_o value.
- If the p-value of a test is greater than α, a $(1-\alpha)*100$ % CI does contain the H_o value.

Example: High efficiency washing machines use less water and less detergent than standard washing machines. Everyone varies in the amount of detergent that they put into a washing machine. Some people might put a little bit more, while others might use a little bit less. The data is below.

$$87 \quad 88 \quad 103 \quad 98 \quad 94 \quad 99 \quad 89 \quad 91 \quad 87 \quad 91$$

a)   The 150 ounce bottle of detergent claims that you get 96 loads– is the population mean significantly different from that value?

b) We had constructed the following CIs for the mean using that same sample. How do they relate to the results of the significance test?

99% CI: (86.95, 98.45)

95% CI: (88.69, 96.70)

90% CI: (89.46, 95.95)

Note: The results of the CI and significance test on the same data will always agree -- as long as the test is two-sided. For one-sided tests, you may need to double the p-value to see the agreement.

Example: Earlier, we tried to determine if the average salary of new graduates was higher than 30,000. The sample mean was $31,000, and the standard deviation was 4,000. (n=50)

a) Make a 95% CI for μ and interpret.

b) We tested before: $H_o: \mu = 30,000$ vs. $H_a: \mu > 30,000$
$$\text{TS: } z = 1.77 \qquad \text{p-value} = 0.0384$$
Does this agree with the 95% CI?

c) What would be the p-value for a two-sided test -- one that asked if the average salary was different from 30,000 dollars?

d) Does this test agree with the 95% CI?

What Effect Does the Sample Size Have on P-values?

Example: A new automobile company claims that the fuel efficiency of a new hybrid car is above 50 miles per gallon. A consumer watchdog group selects 10 cars of this new type. For this sample, they find that the average miles per gallon is 50.3 with standard deviation equal to 2.3.

a)   In this case, Ho: $\mu = 50$ versus Ha: $\mu > 50$. The test statistic is 0.41 with a p-value equal to 0.3448. What conclusion do you make?

b)   The 95% confidence interval is (48.655, 51.945). What interpretation do you make?

c)   In practical terms, is this a problem?

d)   With the same value of sample mean and the sample standard deviation, what would the p-value be if the sample size was 2000?

e)   What would be the 95% confidence interval?

f)   What was the effect of the increasing the sample size?

Mark the following statement as true or false. Explain your reasoning.

"Statistical Significance is the same as practical significance."

Example: Suppose that we had two independent researchers testing to see if a river was contaminated by testing the average pH of the river. A neutral river has a pH of 7. They both use the same set of hypotheses: Ho: $\mu = 7$ versus Ha: $\mu \neq 7$. The first researcher got a p-value of 0.049. The second researcher got a p-value of 0.051. At the 0.05 significance level, what decisions do you make for the two researchers?

What is wrong with using alpha as a strict cutoff point?

9.4 Decisions and Types of Errors in Significance Tests

Samples vs. Populations: When making a decision to either "Reject" or "Fail to Reject" the null hypothesis, there is always the possibility that we are making a mistake, since we are basing the decision on a sample, and not the whole population.

Two Possible Types of Error:
- Type I error –reject H_o when H_o is really true.
- Type II error –failing to reject H_o when H_o is not true.

Significance Level, α
- The level of significance, α, is our willingness to make an error in our conclusion about an experiment. It is not about any error we may make in the experiment (such as measurement error), but the chance we make a mistake in our conclusions due to sampling variability.
- $\alpha = P(\text{Type I Error})$
- The most common level of significance is 0.05, but we also use 0.01 and 0.10.

Which Error Did We Make?
In reality, we never know if we make a Type I or Type II error because we never know everything about the population that we are studying. All we can do is control the probability of making one of these errors.

Decision Table: Fill in the table below with the type of error made.

"The Reality" about the population.	Decision made by using information from the sample.	
	Fail to Reject H_o	**Reject H_o**
H_o is true		
H_o is false		

Examples: Answer the questions for each of the problems below.

1. Suppose that your friend was suddenly accused of murder, and given a jury trial. In the American legal system, you are assumed innocent until proven guilty.

   a) What are the Type I and Type II errors for this situation?

   b) What would happen if we lowered the probability of a Type I error to zero?

   c) As you decrease the level of significance, what happens to the probability of a Type II error?

2. An environmentalist is interested in determining if the pH of the creek water behind his house is affected by the new development upstream. He knows that a neutral stream has a pH of 7. He takes 16 samples of water and determines the pH. Is the pH of the creek significantly different from neutral? What are the Type I and Type II errors that can be made here?

3. A fellow student says that the average salary of graduates in your major is 30,000 dollars per year. You don't think that this is correct. You think that the average salary is higher than this. To show that your friend is wrong, you take a simple random sample of fifty graduates who have graduated in the past 5 years and ask for the amount of their starting salary. You find that the sample mean is 31,000 and the sample standard deviation is 4,000. Is there evidence to support your claim that the population mean starting salary is greater than 30,000?
What are the Type I and Type II errors that can be made here?

9.5 Limitations of Significance Tests Intervals and Significance Tests

- The data must be randomly selected from the population of interest. Bad data will give you useless results.

- Other sampling designs (stratified, clusters) require more complicated hypothesis tests and confidence intervals.

- Outliers can cause the mean to not be representative of the center of the distribution, so all data sets should be checked for outliers. All outliers should be investigated and their effect on the conclusion determined. It may be necessary to use statistical procedures that are resistant to outliers.

- "Do not reject H_o" does not mean "Accept H_o."

- Statistically significant does not necessarily mean practically significant.

- The p-value does not tell us the probability that our assumed hypothesis, H_o, is true.

- It is a bad idea for newspapers to only print statistically significant results.

- True effects may not be as large as initial estimates reported by the media.

- Some tests may be statistically significant just by chance.

- The margin of error of a CI takes into account sampling variability **not** errors in measuring or problems with missing data.

- Significance tests tell us how likely the null hypothesis about a parameter is, whereas a confidence interval gives us a set of values that we believe may contain the population parameter.

Chapter 10 Comparing Two Groups

In Chapters 8 and 9, we learned how to make statistical inferences about population means and population proportions. Now we will learn how to compare two groups.

Explanatory and Response Variables: When comparing two groups, we can think of the problem as a bivariate analysis, as we saw back in Chapter 3.

- **Explanatory variable** is binary – it forms the two "groups" we want to compare.

- **Response variable** can be:
    - quantitative – comparing the means of two treatments

    - categorical – comparing two proportions of successes

Independent vs. Dependent Samples: Depending on how the data is collected, the procedures used to analyze it will be different.
- **Dependent Samples**: Two treatments are given either to the same experimental unit, or two very similar ones.
  Examples:

- **Independent Samples**: Experimental units are randomly allocated to receive one of the two treatments, or selected at random from two populations.
  Examples:

10.1 Categorical Response: Comparing Two Proportions

In this section, we are going to discuss how to compare categorical data from two groups. We want to know the difference between two population proportions. Experimental units are randomly allocated to receive one of the two treatments, or selected at random from two populations, that is, we have two independent samples. Examples:

You can perform a significance test to make statements about the difference between two population proportions and a confidence interval to estimate their difference.

Confidence Intervals for Comparing Two Independent Proportions
All confidence intervals are of the form: estimator ± margin of error

- **Parameter** we want to estimate: $p_1 - p_2$.
- **Estimator** of difference between the two population proportions is $\hat{p}_1 - \hat{p}_2$.
- The margin of error depends on two things: the confidence level we want, and the standard error of the estimator.
- **Sampling distribution** of our estimator is the standard normal, z, so this is the table we use to find the value that will correspond to the desired confidence level.

Deriving the Standard Error
- Standard error for comparing two statistics is: $\sqrt{se(estimate1)^2 + se(estimate2)^2}$

- **Standard error** for comparing two independent proportions is: $\sqrt{\dfrac{\hat{p}_1\left(1-\hat{p}_1\right)}{n_1} + \dfrac{\hat{p}_2\left(1-\hat{p}_2\right)}{n_2}}$

Confidence Interval for $p_1 - p_2$ is: $\left(\hat{p}_1 - \hat{p}_2\right) \pm z\sqrt{\dfrac{\hat{p}_1(1-\hat{p}_1)}{n_1} + \dfrac{\hat{p}_2(1-\hat{p}_2)}{n_2}}$

Assumptions: In order for the resulting confidence interval to be valid, you must have at least 10 successes and 10 failures in the sample.

Interpreting Confidence Intervals for Comparing Two Proportions
- **(- , +):** If the confidence interval includes zero, then $(p_1 - p_2) = 0$ is plausible. There is no statistically significant evidence of a significant difference between the two proportions from the two independent populations.

- **(+, +):** If the confidence interval does not include zero, and the values in the interval are positive **(a, b)**, then p_1 is between **a** more and **b** more than p_2.

- **(-,-):** If the confidence interval does not include zero, and the values in the interval are negative **(-a, -b)**, then p_1 is between **b** less and **a** less than p_2. In other words, p_2 is between **a** more and **b** more than p_1.

Significance Tests for Comparing Two Independent Proportions

1.   **Assumptions**
     •

     •

     •

2.   **Hypothesis**
     •   **Null Hypothesis,** H_o

     •   **Alternative Hypothesis,** H_a

3.   **Test Statistic**, TS
     z-score, summarizes the data

4.   **P-value:** probability that the test statistic equals the observed value or a value more extreme if the null hypothesis is true.

5.   **Conclusion:**
     Small p-values support H_a, so **Reject H_o**
     Large p-values suggest **Failing to Reject H_o** (never accept H_o)

A note about small samples: If the sample sizes are not large enough and a one-sided test is needed, the above procedure for the significance test is not appropriate.

Example: Has there been a significant change in the opinions about affirmative action regarding women in the past decade? The GSS presented people with the following statement: "Because of past discrimination, employers should make special efforts to hire and promote qualified women." In 2000(group 1), 514 out of 791 respondents said they "agreed" or "strongly agreed." We will call this group 1. In 2010, 458 out of 696 respondents said the same. We will call this group 2. Is there a significant difference between the proportion of Americans that agreed or strongly agreed in 2000 and 2010?

a) Create a 95% confidence interval to answer the above question and interpret.

b) Conduct a significance test to determine if there is a significant difference in the proportion of people that agreed or strongly agreed between the 2000 and 2010.

c) Do your results to parts a and b agree?

d) Below is the Minitab output for the confidence interval to compare the proportions from different years from the previous example, the year 1996 (group 1) and the year 2010 (group 2). Interpret.

```
Test and CI for Two Proportions

Sample    X     N  Sample p
1       797  1447  0.550795
2       458   696  0.658046


Difference = p (1) - p (2)
Estimate for difference:  -0.107251
95% CI for difference:  (-0.150827, -0.0636759)
```

10.2 Quantitative Response: Comparing Two Means

In this section, we are going to discuss how to compare quantitative data from two groups. We want to know the difference between the two population means. Experimental units are randomly allocated to receive one of the two treatments, or selected at random from two populations, so we have two independent groups.
Examples:

Confidence Intervals for Comparing Two Independent Means
- **Parameter** we want to estimate: $\mu_1 - \mu_2$.
- **Estimator** of the difference between two population means: $\bar{x}_1 - \bar{x}_2$
- **Standard error** for the difference of two independent means:

$$\sqrt{se(estimate1)^2 + se(estimate2)^2} = \sqrt{\frac{s_1^2}{n_1} + \frac{s_2^2}{n_2}}$$

- **Sampling distribution:** If both populations are normal, the distribution of the statistic is approximately a t distribution with very messy df (they don't even have to be an integer). Minitab can do this automatically.   When doing problems by hand, we will use a **t distribution** with:
  - df = smallest of (n_1-1) and (n_2-1); this is a conservative ("safe") number of df OR
  - df = $n_1 + n_2 - 2$ if the standard deviations and the sample sizes are similar for both groups.
- **Confidence Interval for Comparing Two Population Means:**

- **Assumptions**: We need to assume independent random samples from two groups and that the data from each sample comes from a normal distribution (Normality is only important with small samples.)
- **Interpretations** are similar to the case of the difference of two independent population proportions.

Significance Test for Comparing Two Independent Means
1. **Assumptions**
   -
   -
   -

2. **Hypothesis**
   - **Null Hypothesis**, H_o

   - **Alternative Hypothesis,** H_a

3. **Test Statistic**, TS
   t-score, summarizes the data

4. **P-value:** probability that the test statistic equals the observed value or a value even more extreme if H_o is true.

5. **Conclusion:**
   Small p-values support H_a, so **Reject H_o**
   Large p-values suggest **Failing to Reject H_o** (never accept H_o)

Example: Do men in the US have more time to relax then women in the US? In 2010, the General Social Survey asked participants how many hours a day they had to relax. Data for men and women appear in the table below. Use this information to determine if the population average number of hours spent relaxing per day for men is more than for women.

	Males (Group 1)	Females (Group 2)
Sample Mean	4.04	3.38
Sample Standard Deviation	2.91	2.47
Sample Size	2225	2388

a) Create a 95% confidence interval to answer the above question and interpret.

b) Conduct a significance test to determine if men in the population relax more on average.

Example: Does the amount of time that Americans work per week decrease as they get older and closer to retirement or do older people work more hours in order to save more money to retire? In 2010, the General Social Survey asked participants how many hours they usually work at all of their jobs. Data for those between the ages of 25-35 and 55-65 appear in the table below. Use this information to determine if the average number of hours worked for the younger age group is different from the older age group in the population.

Younger age group (age 25 – 35): 40, 40, 40, 50
Older age group (age 55 – 65) : 40, 40, 40, 40, 50, 50, 50, 70

a) What are the necessary assumptions to conduct a significance test to see if there is a significant difference in the average hours worked per week for the two age groups?

b) Look at the Minitab output below. Interpret.

Two-Sample T-Test and CI: ages 25-35, ages 55-65 w/outlier

```
Two-sample T for ages 25-35 vs ages 55-65
            N    Mean   StDev   SE Mean
ages 25-35  4   42.50    5.00     2.5
ages 55-65  8   47.5    10.4      3.7

Difference = mu (ages 25-35) - mu (ages 55-65)
Estimate for difference:  -5.00
95% CI for difference:  (-15.03, 5.03)
T-Test of difference = 0 (vs not =): T-Value = -1.13  P-Value = 0.288  DF = 9
```

c) The largest data point was removed from the older age group. How does this affect your conclusion above?

```
Two-Sample T-Test and CI: ages 25-35, ages 55-65 w/o outlier

Two-sample T for ages 25-35 vs ages 55-65

             N    Mean   StDev   SE Mean
ages 25-35   4   42.50    5.00     2.5
ages 55-65   7   44.29    5.35     2.0


Difference = mu (ages 25-35) - mu (ages 55-65)
Estimate for difference:  -1.79
95% CI for difference:  (-9.65, 6.08)
T-Test of difference = 0 (vs not =): T-Value = -0.56  P-Value = 0.599  DF = 6
```

(Skipping Ahead To....)

10.4 Analyzing Dependent Samples

Comparing the Means of Dependent Samples

Two treatments are given either to the same experimental unit, or two very similar ones, so we have dependent samples. In other words, the observations are matched pairs.

- **Differences**: First, compute differences between the two responses for each experimental unit (unless specified otherwise, we always assume it was done treatment 1 - treatment 2).

- **Summarize the distribution of the differences**: find the average of these observed differences, $\bar{x}_d$, and their standard deviation, s_d.

- **Parameter** of interest: μ (or μ_d)

- **Estimator** of the mean population difference: $\bar{x}_d$

- **Standard error** of the statistic: $\dfrac{s_d}{\sqrt{n}}$ where n = number of differences

- **Sampling distribution** of the statistic: If the differences are normally distributed, our statistic has a t distribution with n -1 df

Confidence interval for estimating the difference between the population means with dependent samples:

Significance tests for comparing the population means with dependent samples:

- What would be the three possible combinations of alternative and null hypothesis for matched pairs test?

- What would be the test statistic?

- What would be the assumptions?

Example: Do yoga techniques help runners run faster? Ten people who were a part of the town's running club decided to perform a test to see if yoga could really help improve someone's running speed. Each person participated in a 5K run around the same track and each individual was timed. For the next three months, each individual performed the same series of yoga moves three times a week. The participants then ran the same 5K run again in similar conditions. Below are the run times, in minutes, for the 5K runs before and after yoga.

runner	1	2	3	4	5	6	7	8	9	10
before yoga	24	25	30	37	29	30	27	22	26	20
after yoga	21	22	31	33	32	27	26	21	24	19
difference										

a) Find the mean and standard deviation of the differences.

b)   Make a quick plot of the differences. Is it appropriate to use the t procedures?

c)   Make a 95% CI μ and interpret.

d)   Conduct the significance test and interpret the results.

e) Do the results of the confidence interval and significance test agree?

f) What assumptions do we need to make in order for our conclusions to be valid? Are they likely to be satisfied?

Example: Another member of the running club felt that consuming a power bar made with cherry concentrate before running would improve the time on the 5K run. So, this time 8 runners ran the 5K distances after eating a power bar with the cherry concentrate and also after eating a power bar without the cherry concentrate.

runner	1	2	3	4	5	6	7	8
with	21	25	26	30	22	25	30	24
without	20	25	25	27	21	24	32	21
difference								

a) Can you offer the running club some pointers about their experiment?

b) Look at the Minitab output below to determine if the t procedure is appropriate?

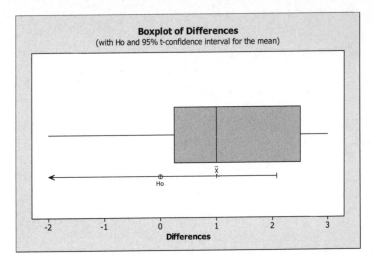

c) There are two Minitab outputs below. Choose the correct one for this problem, and use it to determine if the power bar made with cherry concentrate improved the running speed. Write a conclusion.

Paired T-Test and CI: with, without

Paired T for with - without

	N	Mean	StDev	SE Mean
with	8	25.3750	3.2923	1.1640
without	8	24.3750	3.9256	1.3879
Difference	8	1.00000	1.60357	0.56695

95% upper bound for mean difference: 2.07413
T-Test of mean difference = 0 (vs < 0): T-Value = 1.76   P-Value = 0.939

Paired T-Test and CI: with, without

Paired T for with - without

	N	Mean	StDev	SE Mean
with	8	25.3750	3.2923	1.1640
without	8	24.3750	3.9256	1.3879
Difference	8	1.00000	1.60357	0.56695

95% lower bound for mean difference: -0.07413
T-Test of mean difference = 0 (vs > 0): T-Value = 1.76   P-Value = 0.061

d) How far can we extend the conclusions?

Comparing Proportions with Dependent Samples
When comparing proportions, the response is binary (yes, no) rather than continuous as in the matched pairs test. Each participant is asked two questions, so there are four possibilities for each participant's responses: YY, YN, NN, NY.

Examples:

McNemar's Test for Comparing Proportions with Dependent Samples
- **Hypothesis**: $H_o : p_1 - p_2 = 0$ versus $H_a : p_1 - p_2 \neq 0$
- **Insight:** If there is no difference between $\hat{p}_1$ and $\hat{p}_2$, the number of subjects that have two successes (YY) is equal to the number of subjects that have two failures (NN). Also, the number of subjects with response YN will be equal to the number of subjects with NY.
- **Test Statistic:**

- **Sampling distribution** of the statistic is approximately normal, so you can find the two sided p-value from the z table.

Example: People's opinions about abortion vary widely. Some people feel that abortion should be legal for any reason, others feel that it should be allowed only in special cases, while others don't think that it should be allowed at all. In 2010, the General Social Survey included questions asking people if they thought that a woman should be allowed to have a legal abortion if she was pregnant as a result of a rape or if her health was in danger. Use McNemar's test to see if there is a difference in the proportion of Americans who think women should be able to get a legal abortion if she is pregnant as a result of a rape and the proportion of Americans who think that a woman should be able to get a legal abortion if her own health is seriously endangered?

	Yes, if pregnant because of a rape	No, even if pregnant as a result of a rape
Yes, if her health was endanger	930	117
No, even if her health was endanger	40	124

a) What is the proportion of participants who believe that a woman should be able to have a legal abortion if her health is endanger?

b) What is the proportion of participants who believe that a woman should be able to have a legal abortion if she is pregnant as a result of a rape?

c) State the hypotheses for the McNemar test.

d) Find the test statistic, p-value, and make a conclusion.

(Now back to the section we skipped…..)

10.3 Other Ways of Comparing Means and Comparing Proportions

Relative Risk
Ratios: Instead of comparing two proportions by finding their difference, we can also compare them using ratios. This is quite common with medical studies.

Relative Risk: when one of the outcomes is considered undesirable, the ratio of two proportions $\hat{p}_1/\hat{p}_2$ is called a relative risk.

Example: Suppose that a study followed 2,000 middle age men for ten years. Out of the 1,000 middle age men that were non-smokers, 6 developed lung cancer. Out of the other 1,000 middle age men that were smokers, 92 developed lung cancer.

a)   Find the relative risk when the smokers are group 1 and the non-smokers are group 2. Interpret.

b)   Find the relative risk when the non-smokers are group 1 and smokers are is group 2. Interpret.

Interpreting Confidence Intervals of Relative Risks

- If $p_1-p_2 = 0$, what does the population relative risk equal?

- A confidence interval can be used to estimate the population relative risk, but the formula is too complex for this class.
- To interpret a confidence interval for relative risk, check to see if 1 is included in the interval. If it is, then there is no significant difference between the two groups.

Example: For the lung cancer, software reports that the 95% confidence interval is (6.745, 34.86). What does this mean?

Examples: For the two problems below, identify the two samples and determine if they are dependent or independent. If they are independent, find the relative risk and interpret. If they are dependent, perform McNemar's test to see if there is a difference between the two population proportions.

1. A newspaper article recently reported that men are happier in marriage than women. Is this the view held by the general population? The General Social Survey in 2010 asked 548 women and 502 men if they were very happy in their marriage. Out of these samples, 349 women and 312 men said they were very happy.

	Male	Female
Very Happy	312	349
Not Very Happy	190	199
Total	502	548

2. Many high school students struggle with word problems in mathematics. A new program was designed to help students master word problems using group assignments. Before the program began and after the program was complete, one thousand randomly selected students in the new program were asked to state if they were confident or not confident when they were working out word problems in mathematics. Is there evidence to show that the population proportion of students who were confident when working out word problems before the program is different from the population proportion of students who were confident when working out word problems after the program?

	Confident, after the program	Not Confident, after the program
Confident, before the program	234	203
Not Confident, before the program	243	320

Identifying the type of problem and reading Minitab output.
In Chapters 8, 9 and 10, we have seen how to do statistical inference about different parameters.   In real life, the hardest part is to determine which kind of problem it is, so you can then give Minitab the right instructions, or use the correct formulas.

Examples: For each of the following examples, first determine which kind of problem it is:
- one mean
- comparing means of dependent samples
- comparing two independent means
- one proportion
- comparing proportions with dependent samples
- comparing two independent proportions

Then interpret the results of the Minitab output that are given after the sixth example. Note that for McNemar's test, you will have to analyze the data by hand, since Minitab does not do this test.

1. A class of mostly engineering students was asked to participate in a study about relaxation and heart rates. Each of the students was asked to count their heart beats for one minute. The class then listened to five minutes of relaxing music, and counted their heart beats again. Did the number of heart beats decrease after listening to five minutes of relaxing music?

a) Which kind of problem is this?

b) Conduct the statistical analysis and interpret the results.

2. A department store is trying to sell more of a highly profitable item. They create a new elaborate display with eye-catching colors. After they set up the display, they sell an average of 110 products over the course of 35 days with sample standard deviation of 14. Before the display, they sold 100 products a day. Does it look like the new display has increased sales?

a) Which kind of problem is this?

b) Conduct the statistical analysis and interpret the results.

3. Abortion rights are always a hot topic. Is there a difference in the proportion of men and women that believe abortion is o.k. if the woman gets pregnant as the result of a rape? The General Social survey interviewed respondents about their views about abortion in 2010. For that year, 495 women out of 653 said yes it was o.k. to have an abortion in this situation while 485 men out of 586 men said yes it was o.k. to have an abortion in this situation.

a) Which kind of problem is this?

b) Conduct the statistical analysis and interpret the results.

4. People in the United States have many different opinions about abortion. Some people's views about abortion vary based on the circumstances of the pregnancy. In 2010, the General Social Survey included many questions about people's opinions about abortion in various circumstances. One of the questions asked if it was o.k. for the woman to get an abortion if she was not married and did not want to marry the man; whereas another question asked if it was o.k. for her to get an abortion if she was married, but did not want any more children. In 2010, both of these questions were asked to respondents. Their responses are in the table below. Conduct a test to determine if the proportion of Americans that believe it is o.k. for a woman to get an abortion if she is married, but doesn't want children is different from the proportion of Americans that believe it is o.k. for a woman to get an abortion if she is not married and doesn't want to marry the man.

	Yes, if she is married, but doesn't want more children	No, if she is married, but doesn't want more children
Yes, if she is single and doesn't want to marry the man	479	33
No, if she is single and doesn't want to marry the man	92	599

a) Which kind of problem is this?

b) Conduct the statistical analysis and interpret the results.

5. In the state of Florida, all amendments to the state constitution have to be approved by Florida voters. At least 60 percent of the voters have to approve the amendment. In the upcoming election, a new amendment about education is on the ballot. An exit poll shows that 424 out of 700 people would approve the amendment if the election were today. Is this significant evidence to suggest that the amendment would pass?

a) Which kind of problem is this?

b) Conduct the statistical analysis and interpret the results.

6. Who spends more time answering and sending emails per week, men or women? In 2010, the General Social Survey included a question that asked participants how many hours a week they spent answering or sending emails. The data collected from this question is below. Does the data provide sufficient evidence to see if there is a difference in the amount of time that men and women spend dealing with email?

	Men	Women
Mean	5.70	6.07
Standard Deviation	9.16	9.36
Sample Size	474	629

a) Which kind of problem is this?

b) Conduct the statistical analysis and interpret the results.

Paired T-Test and CI: Before, After
Paired T for Before - After

	N	Mean	StDev	SE Mean
Before	248	71.2298	11.0145	0.6994
After	248	67.2903	11.0807	0.7036
Difference	248	3.93952	7.47630	0.47475

95% CI for mean difference: (3.00445, 4.87458)
T-Test of mean difference = 0 (vs not = 0): T-Value = 8.30 P-Value = 0.000
T-Test of mean difference = 0 (vs > 0): T-Value = 8.30 P-Value = 0.000

One-Sample T
Test of mu = 100 vs not = 100

N	Mean	StDev	SE Mean	95% CI	T	P
35	110.000	14.000	2.366	(105.191, 114.809)	4.23	0.000

One-Sample T
Test of mu = 100 vs > 100

N	Mean	StDev	SE Mean	95% Lower Bound	T	P
35	110.000	14.000	2.366	105.999	4.23	0.000

Test and CI for Two Proportions

Sample	X	N	Sample p
1	495	653	0.758040
2	485	586	0.827645

Difference = p (1) - p (2)
Estimate for difference: -0.0696052
95% CI for difference: (-0.114484, -0.0247264)
Test for difference = 0 (vs not = 0): Z = -3.01 P-Value = 0.003

Test of p = 0.6 vs p > 0.6

Sample	X	N	Sample p	95% Lower Bound	Z-Value	P-Value
1	424	700	0.605714	0.575332	0.31	0.379

Test and CI for One Proportion

Sample	X	N	Sample p	95% CI
1	424	700	0.605714	(0.569512, 0.641917)

Two-Sample T-Test and CI

Sample	N	Mean	StDev	SE Mean
1	474	5.70	9.16	0.42
2	629	6.07	9.36	0.37

Difference = mu (1) - mu (2)
Estimate for difference: -0.370
95% CI for difference: (-1.474, 0.734)
T-Test of difference = 0 (vs not =): T-Value= -0.66 P-Value= 0.511 DF = 1029

This page intentionally left blank.

Name: _____

Student ID# _____

Activity 1: Introduction to the General Social Survey

Purpose: To learn how to use the General Social Survey website to get data on people's opinions on a wide variety of topics.

Website: http://sda.berkeley.edu/GSS/ - click on "GSS - with 'no weight' as the default weight selection"

According to their website, "The GSS is an almost annual, "omnibus," personal interview survey of U.S. households conducted by the National Opinion Research Center. The first survey took place in 1972 and since then more than 38,000 respondents have answered over 3,260 different questions. The mission of the GSS is to make timely, high-quality, scientifically relevant data available to the social science research community." There is a lot of information at this site, and many different ways to access that information. Here we will explore three different ways.

1. **Draft.** We will analyze attitudes towards the drafting of college students into military service.

    a) On the list on the left side look for **Military Issues** and expand the list, then expand **The Draft**.

    b) How many different questions appear under the subject "Draft"?

    c) What is the codename (mnemonic) of the variable that asks if college students should be exempt from the draft?

    d) What is the codename (mnemonic) of the variable that asks if we should return to the draft?

    e) Enter this codename on the box where it says Row and on the box next to Control type "year". Then click on Run the Table at the bottom. Look at the results and scroll down to see the answers by year. Which years was this question asked?

    f) What was the total number of respondents who answered yes, we should return to the draft in 2006?

    g) What was the percentage of respondents who answered yes, we should return to the draft in 2006?

2. **Topical Modules.** This list groups the variables by Modules of a particular topic, including the year the questions were asked. We will analyze Adult Transitions in 2002:

   a) Using the list on the left side of the screen, scroll down to **2002 Topical Module: Adult Transitions** and click on it to open the book.

   b) Type each of the variables below on the box next to Row, then click on Run the Table at the bottom to see the results. Define the following variables:

   finind -

   finind1 -

   ownhh -

   ownhh1 -

   getmar -

   getmar1 -

   c) Of the variables above, which ones are categorical and which ones are quantitative?

3. **Codebooks.** On this list you can quickly find the question you want, if you know the codename used by GSS. Here we will analyze the variable SHOPFOOD:

   a) On the top menu bar select Codebooks, then Standard Codebooks. Select Alphabetical Variable List. Find and select the variable **SHOPFOOD.**

   b) What is the text of the question?

   c) This question was not asked of every respondent – values of 0 indicate they were not asked. Furthermore, this question only applies to people who are "married or living as married." What is the total number of respondents? Of those, how many should really be counted when analyzing the data?

   d) Compute the percentage of respondents who are "married or living as married" who said it is "usually the woman" who shops for groceries in their household using the denominator you determined above?

Activity 2: Describing Data with Dotplots

Purpose: To learn how to describe data sets in terms of their centers, shape, and spread.

Collect data on each of the variables presented on the table below, for each member of your class. Make dotplots of the distribution of each graph on the board. Then, on the spaces provided for each variable, draw a smooth curve to approximately represent the general shape of the distribution, describe the distribution with one or two words, indicate what the center and spread of the distribution are, and whether there are any outliers present.

Variable	Draw shape	Describe shape	Center	Spread	Outliers
height					
number of siblings (not including yourself)					
number of hours it takes to drive "home" from school					
number of hours studied per week					
length of hair (in inches or make some categories?)					

124

This page has been left blank intentionally.

Name: _____

Student ID# _____

Activity 3: Hurricane Season

Purpose: To study the hurricane activity in Florida and the whole Atlantic Basin from 1940 to the present, using a statistical computing package.

Website: http://weather.unisys.com/hurricane/index.html

The data sets used in this activity are downloadable from www.pearsonhighered.com/mathstatsresources

1. **Research hurricane activity.** Visit the Unisys Weather website briefly to familiarize yourself with the data pertaining to hurricane activity in the Atlantic Basin. As you click on each year and then scroll down the page, you get a map with all hurricane and tropical storm activity, then Individual Storm Summary and finally Individual Storm Details.

2. **Get the consolidated hurricane data file** that contains the Individual Storm Summaries from Unisys for the years 1940 – 2010. Copy and paste the data onto your statistical computing package.

3. **Study wind speed and hurricane category:**
    a) Make a histogram of *category*. Describe the shape of the distribution.

    b) According to the histogram, approximately how many hurricanes of each category were there in the years 1940-2010?

category	1	2	3	4	5
number of hurricanes					

    c) Hurricane category is determined by wind speed and pressure. To see the relationship between wind speed and category, construct side-by-side boxplots of the variable *wind,* making one boxplot for each *category* of hurricanes. Describe the plot briefly.

4. Get the Summarized Data file that contains, for each year, a summary of hurricane activity in the Atlantic Basin. For each year, the file has columns for the number of hurricanes of each category, the total number of hurricanes that hit the Atlantic basin that year, the number of hurricanes to hit Florida that year, and how many of those that hit Florida were major hurricanes (FL_major). Copy and paste the data onto your statistical computing package.

5. Study the distribution of number of hurricanes by year:

a) A major hurricane is one that is category 3 or more. The **Summarized Data** file already has counted the major hurricanes to hit Florida, but not for the total Atlantic Basin. Using your statistical computing package, add the number of hurricanes of categories 3, 4, and 5 for each year. Label this column *major*.

b) Make histograms of the variables *total* and *major*. Fill out the table below. Briefly describe the shape, center, and spread of the distribution for each one, including a description of the variables themselves.

	shape	center	spread	Describe the variable
total				
major				

c) Make histograms of the variables *Florida* and *FL_major*. Describe their shapes.

d) Find the mean, median, and mode of the distribution of *Florida* and *FL_major*. **Interpret** these measures of center, including a description of the variables themselves.

Florida		Interpret
	Mean =	
	Median =	
	Mode =	

FL_major		Interpret
	Mean =	
	Median =	
	Mode =	

e) Identify any outliers in the distributions of *total, major, Florida,* or *FL_major.*

Name: _____

Student ID# _____

Activity 4: Exercise and Drinking Habits

Purpose: To determine if there is a relationship between two categorical variables: drinking habits and exercise habits of students.

Can we predict students' exercise habits if we know how much they drink? Here, we will first design a very short questionnaire to determine student's drinking and exercise habits. Then, we will collect data from the students in the class and compute several percentages to see what the data reveals about the relationship between drinking and exercise habits.

1. **Designing the questionnaire – small groups:** Form groups of 3 to 5 students. Each group will determine what they think the definition of "low," "medium," and "high" levels of drinking and exercise are - IN ADDITION to the "no exercise" and "no drinking" categories. Be as specific as you can, i.e. how much, how often, by day, by week. Write your definitions your group agreed on below.

Exercise:
   low: _____

   med: _____

   high: _____

Drinking:
   low: _____

   med: _____

   high: _____

2. **Designing the questionnaire – whole class:** Now each group will discuss their scale with the class, and a common scale for the class will be chosen. Write the definitions the whole class agreed on below.

Exercise:
   low: _____

   med: _____

   high: _____

Drinking:
   low: _____

   med: _____

   high: _____

3. **Collecting the data:** Ideally, we would select students at random from the population. (If this is not feasible, just collect the data for students in your class, understanding that your results will not generalize to a larger population.) To preserve confidentiality, each student should write on a piece of scrap paper what their exercise and drinking habits are (none, low, med, high), according to the definitions above. The results for the class will be tabulated and summarized on the table on the following page.

128

Results Summary:

Exercise	Drinking			
	none	low	med	high
none				
low				
med				
high				

4. **Find the percentages:**

	of non drinkers that engage in:	of low drinkers that engage in:	of med drinkers that engage in:	of high drinkers that engage in:
no exercise				
low exercise				
med exercise				
high exercise				

5. **Describe the results.** Does there seem to be a pattern to this data? If so, briefly explain what it is and how strong it seems.

Name: _____

Student ID# _____

Activity 5: Understanding Correlation

Purpose: To investigate the impact of outliers on correlation and to explore the other concepts of correlation through the use of statistical applets.

Applet: The applet used in this activity, Correlation by Eye, appears on the CD that accompanies the textbook: Agresti/Franklin, *The Art and Science of Learning from Data*, 3e.

1. **Facts about correlation**. Answer the following questions about correlation (r).
   a) What is the strongest the correlation can ever be? _____

   b) If there is no relationship, r is equal to _____.

   c) The correlation coefficient ranges from _____ to _____.

   d) If the points fall in an almost perfect, negative linear pattern, r is close to: _____

   e) If the points fall in an almost perfect, positive linear pattern, r is close to: _____

2. **Determining the value of r by eye.**
   Access the Correlation by Eye statistical applet. A scatterplot will appear. Enter your guess into the box next to Guess. Click on the link "Show r!" The applet will then tell you the real value of r for this data set. Click on "New Data" to get a new data set. Enter a new guess and see how it compares to the real value of r for the data set. Continue guessing until you can find the value of the correlation fairly well.

3. **Let's explore the impact of outliers on r.** To do this, we are going to explore the relationship between someone's current age and the age at which someone got a cell phone. Collect the current age and age at which someone acquired a cell phone from ten student volunteers. Create a scatterplot on the board.
   a) Describe the relationship.

   b) Approximate the correlation r = _____

   c) Enter the data for the ten volunteers into a statistical software package.

   d) Compute the correlation coefficient, r. _____

   e) Suppose that the instructor for the course is 40 years old and first purchased a cell phone when she was 40 years old. Add this data point to the scatterplot. What is the new value for r? r = _____

   f) Find out how an outlier can affect the scatter plot of the data by having your statistical software program create a new scatterplot. How did the plot change from the plot on the board?

4. How does each point affect the correlation?

a) Go back to the applet used in part 2. Erase all of the data in the scatterplot by clicking on the trash can. **Add five points** in an oval shape to the graph in a pattern similar to the plot labeled "a" below. Make sure that the oval is inside the box formed by points (40,40), (40,50), (50, 40), and (50, 50). Click on "Show r." This will give you the value of the correlation. Ignore the message "Not a #!" given by the applet. Mark this in the table below.

b) Then add **two points in the middle of the oval** (as shown on plot "b") and note what happens to the correlation coefficient on the box below the plot. Is r now stronger or weaker than on the previous part?

c) For plot "c," **add one point** at about (45, 60), and note what happens to r.

d) For plot "d," **add one point** at about (60, 55), and note what happens to r.

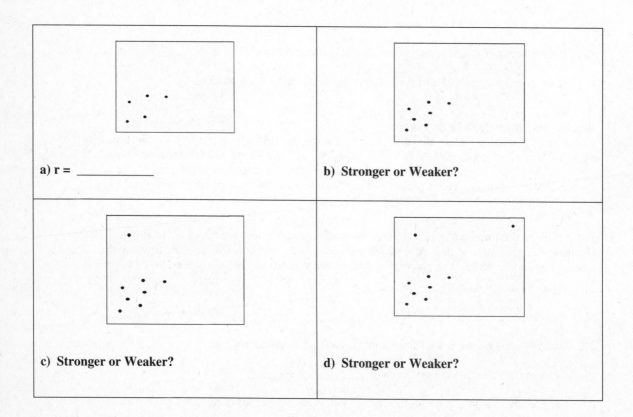

a) r = _____   b) **Stronger or Weaker?**

c) **Stronger or Weaker?**   d) **Stronger or Weaker?**

5. Conclusion. Do outliers have an impact on the correlation coefficient, r? Explain.

131

Name: _____

Student ID# _____

Activity 6: Foot Size, Forearm, and Height

Purpose: To determine how people's height, the length of their forearms, and the length of their feet relate to each other.

1. **Collect the data**: Each student in the class will measure their forearm, that is, the length between their wrist and the inside of their elbow. Each student will also provide their height (in inches) and the length of their feet in inches, using the shoe-size conversion chart provided at the end of this activity.

    Your forearm: _____   **height:**_____   **foot length:**_____

2. **Enter the data** for the whole class into a statistical software program.

3. **Perform a regression analysis for each pair of variables**. Write down the results on the table, and interpret the slope for each one.

x	y	LSR equation	r	Interpret slope
forearm	foot			
foot	height			
forearm	height			

222

4. **Best predictor.** Which variable is best to use when predicting height – forearm or foot length? Explain.

5. **Intercept.** Why do we NOT interpret the intercepts in these cases?

Shoe Size Chart

Women			Men	
Inches	US size		Inches	US size
8.50	5		9.33	6
8.67	5 1/2		9.50	6 1/2
8.83	6		9.67	7
9.00	6 1/2		9.83	7 1/2
9.17	7		10.00	8
9.33	7 1/2		10.17	8 1/2
9.50	8		10.33	9
9.67	8 1/2		10.50	9 1/2
9.83	9		10.67	10
10.00	9 1/2		10.83	10 1/2
10.17	10		11.00	11
			11.17	11 1/2
			11.33	12
			11.50	12 1/2
			11.67	13

Name: _____

Student ID# _____

Activity 7: Sampling Legos®

Purpose: To study the benefits of simple random sampling over other sampling methods.

For this activity, the class should be divided into groups of three or four students. Each group will be provided with a bag or box containing around 50 Lego® pieces. This container will be your "population."

Lego® pieces come in many shapes and sizes, and each piece has a number of "dots" that are used to connect it to other pieces. Students will try to determine the average number of dots per piece in their population using three different methods: guessing, convenience sampling, and simple random sampling.

The sampling issues that we will explore here with the Legos® are applicable to many sampling situations. You can think of these Lego® pieces as representing apartments in a building, with the number of dots on each Lego® corresponding to the number of people who live in each apartment. Or each Lego® could represent a person, and each dot represents the number of beers that person had last Friday. In these cases we would be interested in estimating the average number of people per apartment, or the average number of beers consumed per person.

Important Notes:
- **DO NOT take the Lego® pieces out of the container until instructed to do so.**
- **Each student will do parts 1, 2, 3c, 3d, and 5 individually. Part 4 will be done by the group sharing the container of Legos®. Part 6 will be done by the whole class together.**

1. **Guessing the population mean.** Without taking any pieces out, or shaking the container, each student should look at the pieces, and guess the average number of dots on the Lego® pieces in the population. This is similar to what we do when we guess at the average number of people per apartment or the average number of beers per person, based only on the people we know.
   Your guess: _____

2. **Conduct a convenience sample.** Look into the container and take out ONLY five Lego® pieces that you think are representative of the population. Count the number of dots on each Lego® piece in your sample, and compute the average. Write your results on the next page. Put the Lego® pieces back in the bag and have the other students in your group select their own sample and compute their own average. Even though this time you are selecting a sample and computing the average, you did not use a mechanical method to select your sample. This is similar to convenience samples where we are more likely to select some people than others because they "stand out" in some way.

	Item 1	Item 2	Item 3	Item 4	Item 5	Average
# dots						

134

3. **Conduct a simple random sample.**
   a) Each group should take all the Lego® pieces out of the container and count the number of Lego® pieces in the population. *N*=_____

   b) As a group, arrange all the Lego® pieces in your population in such a way that you can assign a number to each one. This can be done in a line, a grid, etc.

   c) Each student in the group will pick a random sample of five items from the population. Using a statistical software package, a calculator, or a random number table, select five random numbers from 1 to N. If any of the numbers repeat, select another one until you have five different values. Write the numbers that you generated into the row labeled "Order" below. Each person in the group should have his or her own list of random numbers.

   d) The five random numbers tell you which Lego® pieces to use as your random sample. Using these numbers and the line or grid of pieces you created before, find the pieces, count the number of dots on each piece, and compute the average number of dots for your sample. Replace the pieces in their proper place before the next student in your group selects their sample and computes their own average.

	Item 1	Item 2	Item 3	Item 4	Item 5	Average
Order						
# dots						

4. **Find the population mean.** As a group, now find the mean number of dots on all of the Lego® pieces in the container your group used (your population). μ = _____

5. **Fill in the table** with your results from parts 1, 2, 3 and 4 so you can see your here estimates all together and see how they compare to the true population mean they were trying to estimate.

Your guess at the average	Your convenience sample average	Your simple random sample average	Your population mean

6. **Which sampling method works better?** Compare each of your three estimates to the true population mean. Each student gets to award 1 point to the best method, according to their results. If one of the methods was closer to μ than the others that method gets the whole point (higher or lower does not matter – just how close it was). If two of the methods were equally close, they each get 1/2 point. If all three estimates were exactly the same, each method gets 1/3 of a point.

   Your point: guess_____   convenience sample ____random sample ____

7. **Collect the points for the whole class:**

   Class points: guess_____   convenience sample ____random sample ____

8. **Which method worked better overall?** Is this what you expected? Explain.

Name: _____

Student ID# _____

Activity 8: Should You Trust Everything You Read?

Purpose: To learn how to critically analyze the results of research studies as presented by the media.

1. **Read the news article** selected by your instructor.

2. **Summarize the article** in a few sentences.

3. **Critique the article**
    a) What is the explanatory variable? Is it a categorical or quantitative variable?

    b) What is the response variable? Is it a categorical or quantitative variable?

    c) What would be the best graph to describe the data - a boxplot or a scatterplot?

    d) Describe the sampling method used, if possible. Is this a good method?

    e) List the variables that are controlled in the study.

136

f) List any potential lurking variables.

g) What are the conclusions of the article?

h) Do you agree with the conclusion in the article? Explain your answer giving statistical reasoning in complete sentences.

Name: _____

Student ID# _____

Activity 9: Normal Rent Payment

Purpose: To determine if students' rent payments follow an approximately normal distribution.

1. **Record the rent payments** of the students in the classroom, in dollars.

2. **Make a stemplot** of the payments, either by hand or using a statistical software package, and place it below. Describe the shape of the plot.

3. **Find the mean and standard deviation** of the rent payment values using your calculator or a statistical software package.

    Mean:_____   Standard Deviation:_____

4. **State the Empirical Rule**.

5. Fill the table below to compare the proportion of people in the class whose rent payments fall in the regions specified to the proportions expected in the same region if the distribution were normal.

Region	Region in terms of rent payments	Number of observations in region	Percentage of observations in region	Percentage expected in region, if normal
within one stdev of the mean				
within two stdev of the mean				
within three stdev of the mean				
below the mean				

6. Does the distribution seem approximately normal? Justify your answer using the information found on the table, as well as the shape of the distribution of rent payments in the class.

Name: _____

Student ID# _____

Activity 10: Sampling Distribution of a Proportion

Purpose: To study the sampling distribution of the sample proportion using Java Applets on the Internet.

Applet: The applet used in this activity, Sampling Distributions, appears on the CD that accompanies the textbook: Agresti/Franklin, *The Art and Science of Learning from Data*, 3e.

1. **Terms of the simulation.** In class, we studied the sampling distribution of $\hat{p}$, the sample proportion of successes in a binomial experiment. We saw that this distribution is approximately normal if *np* and *n(1-p)* are both greater than or equal to fifteen. The parent population of the data is binary, meaning that it has only two potential responses: success (1) or failure (0). Using the applet, we are going to repeatedly draw samples from the binary distribution and compute the sample proportion of successes. This will allow us to see when the sampling distribution of $\hat{p}$ can be approximated with the normal distribution. Familiarize yourself with the applet and answer the following questions.

   a) Each of the samples that will be drawn from the parent distribution will be of the same size. What is the symbol that the website uses to represent the size of each sample? _____

   b) What is the symbol used to represent how many times we collect these samples? _____

2. **Simulation.** For each setting of n and p given in the table that follows, compute the values of *np* and *n(1-p)*. Determine if *np* and *n(1-p)* are greater than or equal to fifteen. Then use the applet to see the sampling distribution and determine if the normal approximation is good for each case (get at least one thousand samples for each combination of *n* and *p*). Select the value of p by using the drop down box next to the population graph. For each combination, determine if the graph shows that the sampling distribution of $\hat{p}$ is close to normal. Look at symmetry, continuity (no big gaps in the data), and tails.

140

n	p	np	n(1-p)	both ≥ 15?	Sketch	Continuity? (No big gaps in data)	Symmetry?	Normal Approximation Good?
10	0.70							
30	0.70							
50	0.70							
100	0.70							
10	0.50							
30	0.50							
50	0.50							
100	0.50							
10	0.10							
30	0.10							
50	0.10							
100	0.10							
1000	0.10							

3. **Play with the applet a bit.** Compare the results for each value of p as n increases. Then see what happens for different values of p below and above 0.5.

4. **Summary.** In your own words, explain what combinations of n and p are necessary for the sampling distribution of $\hat{p}$ to be approximately normal, and why.

Name: _____

Student ID# _____

Activity 11: Central Limit Theorem

Purpose: To study the sampling distribution of the statistic $\bar{x}$ (the sample mean), using Java Applets on the Internet.

Applet: The applet used in this activity, Sampling Distributions, appears on the CD that accompanies the textbook: Agresti/Franklin, *The Art and Science of Learning from Data*, 3e.

1. **Terms of simulation.** We are going to use statistical applets to investigate the central limit theorem. The statistical applet will allow us to draw a sample many times from a parent distribution. The means of all of the samples will be computed, so that we can see the shape, mean, and standard deviation of the sampling distribution of $\bar{x}$. Familiarize yourself with the applet and answer the following questions.

   a) Each of the samples that will be drawn from the parent distribution will be of the same size. What is the symbol that the website uses to represent the size of each sample? _____

   b) What is the symbol used to represent how many times we collect these samples? _____

2. **Simulation.** For each parent distribution and sample size given on the table that follows, write down the mean and the standard deviation (given by the computer) in the first column. Then, compute the values of the mean and standard deviation of the distribution of $\bar{x}$ in the theoretical columns using the values that the Central Limit Theorem specifies. Then use the applet to get the distribution (get at least one thousand samples for each case). Record the mean and standard deviation of your simulation in the observed column. Comment on the shape of the graph. For the "**custom**" parent population, use the mouse to create a bimodal distribution. NOTE –When you look at the shape, imagine it being smoother.

Parent Population (given on applet)	Sampling Distribution of $\overline{X}$				
	Sample Size	Compute: Theoretical Mean Stdev		From Simulation: Observed Mean Stdev	Shape
Normal μ= σ=	2				
Normal μ= σ=	30				
Uniform μ= σ=	2				
Uniform μ= σ=	30				
Skewed μ= σ=	2				
Skewed μ= σ=	30				
Custom-Bimodal μ= σ=	2				
Custom-Bimodal μ= σ=	30				

3. **Summary.** Based on the results of the simulation, what happens to:

a) the **shape** of the distribution of $\overline{x}$ as n increases?

b) the **mean** of the distribution of $\overline{x}$ as n increases?

c) the **standard deviation** of the distribution of $\overline{x}$ as n increases?

Name: _____

Student ID# _____

Activity 12: Confidence Interval for a Proportion

Purpose: To find data to answer a question of interest, construct a confidence interval, and make conclusions for a population proportion.

Website: **http://sda.berkeley.edu/GSS/ -** click on "GSS - with 'no weight' as the default weight selection"

1. **Select a research question from the General Social Survey.** We first learned how to use it in Activity 1, so you may want to go back to it if you need a refresher. The question you select must be one that can be summarized with a proportion of successes. Some examples of questions you may use are: FEPRES, FEPOL, FEWORK, HAPPY, SATJOB7.

2. **Write down the question selected.** What will be considered a success?

3. **Summarize the data.**

   number of successes: $X =$

   total number of observations: $n =$

4. **Identify the parameter and the statistic** in symbols and with words. Compute their values if possible. Explain how they are related to each other.

144

5. Assumptions. What are the assumptions necessary for making inferences in this case? Have they been met?

6. Construct a 95% confidence interval for the population proportion.

7. Interpret the results of the confidence interval.

Name: _____

Student ID# _____

Activity 13: CI for the Mean – What Does It Mean?

Purpose: To learn to identify correct and incorrect ways to interpret a confidence interval.

In this activity, we will try to estimate μ, the average number of hours per week all students at your college or university spend studying outside of class. You will collect the data in class to create a 95% confidence interval for μ. Then you will consider the statements presented in part 5, and determine whether they are correct or incorrect interpretations of the confidence interval.

1. **Collect the data**. On a piece of scrap paper, write down how many hours you spend studying outside of class on a typical week. Pass the paper to the front of the class.

2. **Summarize the data**. Using your calculator, or a statistical software package, find the mean and standard deviation for the "study" data collected for the whole class.

    $\overline{x} =$ _____ $s =$ _____ $n =$ _____

3. **Construct a 95% confidence interval for μ.**

4. **Assumptions**. What assumptions need to be satisfied for our interval to be valid? Are they likely to be satisfied? Explain.

146

5. **Interpretations**: Suppose a random sample of 114 students was chosen, and each student was asked how many hours he or she studies each week. The resulting 95% confidence interval for μ was (8.9, 11.8). Determine if each one of the following statements is true with a **capital "T"** or false with a **capital "F."**

_____ a) 95% of all students study between 8.9 and 11.8 hours per week.

_____ b) 95% of all sample means will be between 8.9 and 11.8.

_____ c) 95% of samples will have averages between 8.9 and 11.8.

_____ d) For 95% of all samples, μ will be between 8.9 and 11.8.

_____ e) For 95% of all samples, μ will be included in the resulting 95% confidence interval.

_____ f) The formula produces intervals that capture the sample mean for 95% of all samples.

_____ g) The formula produces intervals that capture the population mean for 95% of all samples.

Name: _____

Student ID# _____

Activity 14: Testing for Proportions

Purpose: To find data to answer a question of interest, conduct a significance test and make conclusions for a population proportion.

Website: http://sda.berkeley.edu/GSS/ - click on "GSS - with 'no weight' as the default weight selection"

1. **Select a research question from the General Social Survey.** We first learned how to use it in Activity 1, so you may want to go back to it if you need a refresher. The question you select must be one that can be summarized with a proportion of successes. Some examples of questions you may use are: FEPRES, FEPOL, FEWORK, HAPPY, SATJOB7.

2. **Write down the question selected.** What will be considered a success?

3. **State the null and alternative hypothesis to be tested:**

   H_o:

   H_a:

4. **Summarize the data.**

   number of successes: $X =$

   total number of observations: $n =$

5. **Identify the parameter and the statistic** in symbols and with words. Compute their values if possible. Explain how they are related to each other.

148

6. **Assumptions.** What are the assumptions necessary for making inferences in this case – for the confidence interval and the significance test? Have they been met?

7. **Conduct a significance test** for the population proportion. What is the value of the test statistic?

8. **P-value.** What is the p-value of the test?

9. **Interpret the results of the test** based on the p-value.

10. **Construct a 95% confidence interval** for the population proportion.

11. **Interpret the results of your interval.**

12. **Agreement.** Do the conclusions of the confidence interval and the significance test agree? Explain.

Name: _____

Student ID# _____

Activity 15: Interpreting P-Values

Purpose: To conduct a significance test and learn to identify the correct and incorrect ways to interpret a p-value.

According to the General Social Survey, the average number of hours of TV watched per day by an American adult was 3.03 hours. Test the claim that students at your college or university spend less than 3.03 hours watching TV per day.

1. **State the null and alternative hypothesis.**

   H_o:

   H_a:

2. **Collect the data**. How many hours do you spend watching TV every day? _____

3. **Assumptions.** What are the assumptions necessary for making inferences in this case? Have they been met?

4. **Conduct a significance test** of the data. What is the value of the test statistic?

5. **P-value.** What is the p-value of the test?

6. **Interpret the results of the test** based on the p-value.

150

7. **Interpretations.** Suppose a random sample of 114 students gave a p-value of 0.0011 for testing the same hypotheses. Determine if each one of the following statements is True or False.

_____ a) We can reject the null hypothesis at $\alpha = .05$, but not at $\alpha = .01$.

_____ b) We can reject the null hypothesis at $\alpha = .10$, but not at $\alpha = .05$.

_____ c) We can reject the null hypothesis at $\alpha = .10$, $\alpha = .05$, and $\alpha = .01$.

_____ d) 99.89% of students in the sample spend less than 2.98 hours/day watching TV.

_____ e) 99.89% of students in the population spend less than 2.98 hours/day watching TV.

_____ f) The probability of observing results as low as these if the alternative hypothesis is true is 0.0011.

_____ g) The probability of observing results as low as these if the null hypothesis is true is 0.0011.

Name: _____

Student ID# _____

Activity 16: Two Scoops of Raisins?

Purpose: To conduct significance test to determine the validity of the Kellogg's® Raisin Bran slogan that there are two scoops of raisins in every box.

Kellogg's® Raisin Bran claims that it puts 2 scoops of raisins in every box. According to their Consumer Affairs Department, for a 15oz box, each of these scoops is a quarter of a cup, so these 2 scoops should then total about half a cup of raisins. One half cup of raisins is about 160 raisins. In this activity we will test their claim to determine if indeed the 15oz boxes contain at least 160 raisins on average vs. the alternative that the boxes contain less than that amount.

1. **State the null and alternative hypothesis** in terms of the average number of raisins that should be in every box of Kellogg's® Raisin Bran of this size.

   H_o: _____ H_a: _____

2. **Collect the data.** The instructor will divide the class into ten groups. Each group is responsible for counting the number of raisins in one of the boxes. Each group will record their answer on the board when they are finished.

3. **Record the data.** Fill in the chart below with the class data.

Box	1	2	3	4	5	6	7	8	9	10
Number of Raisins										

4. **Create a dot plot of the data.**

152

5. **Assumptions.** What are the assumptions necessary for the validity of the significance test of the one
    sample mean case? Are they likely to be satisfied?

6. **Conduct a significance test**. What is the value of the test statistic?

7. **P-value.** What is the p-value for the test?

8. **Decision.** What should be our decision at $\alpha = 0.05$?

9. **Briefly interpret** the results of the significance test. Do you believe the boxes contain 160 raisins, on
    average? Explain.

10. **z vs. t.** Why do we need to use the t table in this problem instead of the z table?

Activity 17: Haircuts – Who Spends More?

Purpose: To conduct a significance test and to make a confidence interval for comparing the means of two groups.

Who spends more on haircuts, men or women? Are you considering only the amount spent on each haircut, or are you taking into account how often men and women tend to go for a haircut? In this activity we will first test the assertion that men's haircuts are cheaper than women's. We will also take into account the frequency of the haircuts, to test if the overall amount spent on haircuts per year is different for the two genders.

We will use the following symbols throughout this activity:
μ_{YM} = mean amount of money spent on haircuts each year by males
μ_{YF} = mean amount of money spent on haircuts each year by females
μ_{HM} = mean amount of money spent on each haircut by males
μ_{HF} = mean amount of money spent on each haircut by females

1. State the null and alternative hypotheses for each of the following questions.

   a) Do women really spend more money per haircut than men, on average?

   H_o: H_a:

   b) Over the course of a year, who spends more on haircuts, males or females?

   H_o: H_a:

2. **Collect the data.** Answer the following questions on a piece of scrap paper and pass it to the front of the classroom.
   a) Are you male or female?
   b) How much did you spend on your last haircut? Include amount paid in tips, but not amount spent on coloring, treatments, or any other services.
   c) How much do you spend on haircuts per year?

3. **Record the data for the whole class on the board.**

4. **Which case.** Should the questions above be answered by conducting inference procedures about difference of two independent means or about matched pairs differences? Explain.

154

5. **Assumptions**. What are the assumptions necessary for making inferences in this case? Are they likely to be satisfied?

6. **Do women spend more than men on an individual haircut?**

    a)   What is the p-value for the significance test? _____

    b)   Interpret the results of the significance test.

    c)   95% CI:

    d)   Interpret the confidence interval.

7. **Is there a difference between the amount of money that men and women spend on haircuts per year?**

    a)   What is the p-value for the significance test? _____

    b)   Interpret the results of the significance test.

    c)   95% CI:

    d)   Interpret the confidence interval.

8. **Conclusions**. Write two or three sentences that discuss your findings about the amount of money that males and females spend on haircuts per an individual haircut and per year.

Name: _____

Student ID# _____

Activity 18: Matched Pairs

Purpose: To learn to make confidence intervals and conduct significance tests for mean of matched pairs differences.

Some people believe that listening to soft music has a calming effect on their bodies. In this activity, students will measure their pulse rate before and after listening to five minutes of soothing music in the classroom, and use this data to test the claim.

1. **Collect the data.**

   a) To measure your pulse rate, you can place your index and middle finger on either the side of your neck or the underside of your wrist, until you feel the blood pumping regularly. When everyone is ready, the instructor will give a signal to start counting the number of pulses. The instructor will then give a signal to stop after exactly one minute (60 seconds). The number of pulses in one minute is your pulse rate. Measure and record your pulse rate before listening to music: _____

   b) The instructor will play five minutes of soothing music. During this time, students should simply relax and listen to the music; no talking, reading, writing, using computers, etc.

   c) Measure and record your pulse rate after listening to music: _____

2. **Record the data** for the whole class on the board. There should be two columns, marked "before" and "after." The data for each student should appear in one row.

3. **Construct a dotplot of the differences** between the "before" and "after" pulse rate. Describe the main features of the plot.

156

4. **Assumptions.** What are the assumptions necessary for making inferences in this case? Are they likely to be satisfied?

5. **Construct a 95% confidence interval** for the mean difference in pulse rate before and after listening to music.

6. **Interpret the results of your interval.** Was there a significant reduction in pulse rate? Explain.

7. **Conduct a significance test** to see if the heart rate decreased, on average.

   H_o: H_a:

   test statistic:

   p-value:

   conclusion:

8. **Agreement.** Do the conclusions of the confidence interval and the significance test agree? Explain.

Name: _____

Student ID# _____

Activity 19 Investigating the Differences Between Groups

Purpose: To write a three question survey and analyze the results of the survey using a confidence interval for comparing the means and the proportions of two groups.

In this class we have seen the results of surveys many times. This time, you are going to write your own three question survey and then analyze the resulting data.

1. **Determine the groups**. First, determine what two independent groups you would like to compare. Think of two groups in which a person could only belong to one. Write a question that asks the respondents to which group do they belong. (For example, "Are you a vegetarian or non-vegetarian?")

2. **Quantitative Question.** Write a question that would allow you to compare quantitative data for the two groups. This will be Topic #1. (For example, "How much did you spend on lunch today?")

3. **Categorical Data.** Write a question that would allow you to compare binary (yes/no) data for the two groups. This will be Topic #2. (For example, "Did you have a soft drink with lunch? Yes or No")

4. **Collect the data from the class.** Answer the above three questions on a piece of scrap paper and pass it to the front of the classroom. Make sure that you label each answer.

5. **Record the data for the whole class on the board.**

158

6. **Analysis for Topic #1.**

   a) **Which case?** Should the question for Topic #1 be answered by conducting inference procedures about difference of two independent means, difference of two independent proportions or about means from dependent samples? Explain.

   b) **Assumptions.** What are the assumptions necessary for making inferences in this case? Are they likely to be satisfied?

   c) **Construct a 95% CI:** _____

   d) **Interpret the confidence interval.**

7. **Analysis for Topic #2.**

   a) **Which case?** Should the question for Topic #2 be answered by conducting inference procedures about difference of two independent means, difference of two independent proportions or about means from dependent samples? Explain.

   b) **Assumptions.** What are the assumptions necessary for making inferences in this case? Are they likely to be satisfied?

   c) **Construct a 95% CI:** _____

   d) **Interpret the confidence interval.**

8. **Conclusions.** Write two or three sentences that discuss your findings from your survey.

Name: _____

Student ID# _____

Activity 20: Contingency Tables

Purpose: To learn how to analyze data from a contingency table.

How proud are Americans of our country's "fair and equal treatment of all groups in society"? Is there a difference among the racial groups on how they would answer this question? The following table was constructed using data from GSS, by tabulating the variable "proudgrp" by race, using data for the most recent year this question was asked.

| Race | How proud are you of America's fair and equal treatment of all groups in society? | | | | |
	Very proud	**Somewhat proud**	**Not very proud**	**Not proud at all**	**Total**
White	244	484	169	40	**937**
Black	26	63	44	20	**153**
Other	23	27	20	2	**72**
Total	**293**	**574**	**233**	**62**	**1162**

1. **Identify the explanatory and response variables** in this table.

2. **Compute the conditional probability** of each answer, by race of the respondent. Enter those percentages on the table, below each count. Should the row or column percentages add up to 100%?

3. **State the null and alternative hypothesis** of the test to determine if the answer to this question is independent of race.

160

4. **Assumptions**. State the assumptions necessary for the validity of this test. Do they seem satisfied? Explain.

5. **Expected count**. What is the expected count for "white" / "very proud" under the null hypothesis?

6. **Compute the test statistic**.

7. **Find the p-value of the test**.

8. **Degrees of freedom**. How many degrees of freedom does the distribution used to find the p-value have?

9. **Interpret the results** of the test.

10. **Strength of association**.
    a) Can the chi-squared test determine if there is a strong association between the variables? Explain.

    b) Find the difference in the proportion of blacks vs. whites that answered "not proud at all" to this question. Interpret this difference.

    c) Find the ratio of the proportions of blacks vs. whites that answered "not proud at all" to this question. Interpret this relative risk.